Computational Surface and Roundness Metrology

Bala Muralikrishnan • Jay Raja

Computational Surface and Roundness Metrology

Springer

Bala Muralikrishnan, PhD
Research Engineer
KT Consulting, Inc.
2545 West 10th Street, Suite A
Antioch, CA 94509
USA

and

Adjunct Assistant Professor
Department of Mechanical Engineering
and Engineering Science
University of North Carolina
at Charlotte
9201 University City Boulevard
Charlotte, NC 28223
USA

Jay Raja, PhD
Professor and Chair
Department of Mechanical Engineering
and Engineering Science
University of North Carolina
at Charlotte
9201 University City Boulevard
Charlotte, NC 28223
USA

ISBN 978-1-84996-773-0 e-ISBN 978-1-84800-297-5

DOI 10.1007/978-1-84800-297-5

British Library Cataloguing in Publication Data
Muralikrishnan, Bala
Computational surface and roundness metrology
1. Surfaces (Technology) – Measurement 2. Roundness measurement
I. Title II. Raja, J.
620.4'4

Cover design: eStudio Calamar S.L., Girona, Spain

Printed on acid-free paper

9 8 7 6 5 4 3 2 1

springer.com

To

Prof. V. Radhakrishnan

INAE Distinguished Professor
College of Engineering, Trivandrum, India
and
Former Dean, Department Head and Professor
Indian Institute of Technology, Chennai, India

Preface

The purpose of this book is to provide insight into computational techniques commonly used in surface and roundness metrology. It is intended to serve as a reference for engineers who routinely use a variety of metrology tools and are interested in understanding the underlying mathematical basis of the data processing from raw data to final parameters. This book can also be used as a text for advanced courses in engineering metrology for students already familiar with measurement methods and practices.

Most of the material is presented with the objective of enabling the reader to not only assimilate the mathematics involved, but also to be able to quickly implement the techniques using a standard math package. For that reason, a number of exercises are included that highlight the implementation of the algorithms using simulated data of surface features. The book employs MATLAB extensively for these exercises. Familiarity with MATLAB will aid in understanding the material.

Also, it should be pointed at the outset that this book assumes a general familiarity with engineering metrology methods; for readers unfamiliar with some of the measurement techniques whose computational aspects are discussed, several excellent textbooks are available and referenced in Chap. 1.

Acknowledgment

N. George Orji carefully reviewed the entire manuscript, offering invaluable comments and suggestions covering both technical and editorial aspects. We are extremely grateful for the generous gift of his time and effort. Numerous colleagues also volunteered to review sections of the text, provided insight into some of the techniques discussed here, and also supplied several excellent references. We express our sincere gratitude to the following people: Christopher Brown, Brian Boudreau, Donald Cohen, David Coleman, Mark Malburg, Abu Mohamed Ali, Eric Novak, Suresh Ramasamy, Jay Ravindran, T. Brian Renegar, Vijay Srinivasan, Eric Stanfield and Theodore Vorburger.

The Center for Precision Metrology at the University of North Carolina at Charlotte offers an extraordinary environment for the education and practice of engineering metrology. The center's incredible intellectual talent, state-of-the-art laboratories, and its web of academic and industrial networks offer students and faculty unique opportunities for research and teaching. This text has benefited greatly from the center's resources and the energy and enthusiasm of its faculty, staff and students. We (the first author as a former graduate, and the second as faculty) thank them for their support.

Acknowledgement

Contents

Part II Advanced Filtering

Part III Fitting

Notation

Filtering and Parameterization

x, y	spatial location of points (arrays)
z	measured surface heights (either array or matrix)
w	waviness profile points
r	roughness profile points in filtering
n	number of profile points (length of array z) for 2D profiles
nx, ny	number of profile points along x and y for 3D topography data sets
m	number of points in the weighting function (2D profile filtering)
mx, my	number of points in the weighting function (3D topography) along x and y
$x(i), y(i)$	ith element of x and y
$z(i, j)$	(i, j)th element of z
zf	Fourier transform of z
$\Delta x, \Delta y$	spacing along x and y in units of length
l	profile length
lx, ly	profile lengths along x and y
S	weighting function (filter in time/spatial domain)
Sf	transmission characteristics (filter in frequency domain)
λ	sinusoidal wavelength
f	sinusoidal frequency
fx, fy	frequencies along x and y
$\lambda c, \lambda f, \lambda s,$	cutoff in units of length for 2D profile filtering
$\lambda cw, \lambda sw$	cutoff in units of length for 2D profile filtering
$\lambda xc, \lambda yc$	cutoff in units of length for 3D topography filtering along x and y
ω	undulations per revolution (UPR)
ωc	cutoff expressed in UPR
E	objective function to minimize or maximize
x_c, y_c, z_c	center coordinates of a least-squares best-fit circle
d	array of deviations at each of the sampling points from best-fit geometry

R	array of radial distances from center of circle to each sampling point
θ	array of angular positions at each sampling point

Fitting

x, y, z	point coordinates in space (each is an array containing n points)
x_c, y_c, z_c	center coordinates of a least-squares best-fit circle
r	radius of least-squares best-fit circle (limaçon)
d	array of deviations at each of the sampling points from best-fit geometry
R	array of radial distances from center of circle to each sampling point (such as in a roundness profile)
n	number of sampling points
E	objective function to minimize or maximize
θ	array of angular positions at each sampling point
λ	Lagrange multiplier

Uncertainty

x, y	spatial location of points (arrays)
z	measured surface heights (either array or matrix)
w	waviness profile points
r	roughness profile points
u	standard uncertainty
X	input quantity, may be an array
Y	output quantity
R	part form error (array containing n points), roundness profile
Rf	Fourier transform of part profile
Sx , Sy	spindle error (array containing n points) along x and y directions
Sfx , Sfy	Fourier transform of spindle errors
θ	array of angular positions at each sampling point
ω	array of frequencies (UPRs)
n	number of sampling points in a roundness trace
Hf	transfer function (in frequency domain)
ϕ	angle between probes in multi-probe or angle between positions in multi-position methods
m_i	the ith measured profile or probe reading (in multiple probes or position methods), an array of n points
m	a combination of measured profiles
mf	Fourier transform of m

Chapter 1
Introduction

1.1 Surface and Roundness Metrology

Surface finish is a crucial link between a component, the manufacturing process that generated it, and the functionality that is expected of it (Whitehouse 1994). This relationship between surface finish, part functionality, and manufacturing process parameters is the primary reason for the measurement, characterization, and study of surface texture.

The science of surface metrology can be traced back to the 1940s, to the seminal work of researchers at the Rank Taylor Hobson Company in UK (Reason *et al.*, 1944). Stylus-based techniques for surface texture measurements have been in use since that time. Reason (1979) and Sherrington and Smith (1988a) survey the progress made in surface topography measurements especially using the stylus method. Optical techniques are also used for surface texture measurement. Sherrington and Smith (1988b) review optical techniques in surface metrology. Roundness measurements were performed in the 1960s and Reason (1966) presents an early report. Whitehouse (1994, 1997, 2004), Thomas (1999), Smith (2002), and Jiang *et al.* (2007) trace the evolution of surface metrology and the different measurement techniques used.

1.2 Scope and Objectives

A surface texture or roundness profile, as acquired by any metrology instrumentation, is a discrete data set. That is, the deviations from the nominal geometry are expressed as a function of spacing coordinates. Although real surfaces are continuous, metrology instrumentation can only acquire a finite digital sample. This finite dimensional array collected by typical commercial instruments is still a fairly large volume of data; a typical surface profile can contain 8,000 to 25,000 data points while a roundness trace can contain 360 to 7,200 points. An area map of a surface can contain 1,024 × 1,024 points or more.

Such a large array or matrix of data is "processed" by software in commercial instrumentation to provide parameters that we are familiar with. These parameters are the link between the component's surface texture, the functionality expected of it, and the manufacturing process that created this surface finish in the first place.

But what does "processing" really mean?

A surface, even after careful alignment with the measurement instrument, may have some residual slope during profile or topography measurement. In some instances, the nominal geometry may be curved, such as when extracting a profile from a surface of a sphere or cylinder. In roundness metrology, the measurement is often superimposed by an eccentricity between the part and the spindle. These measurement setup and part-induced errors can be rectified in software by a process known as *fitting*. Fitting substitute geometry, such as a line, plane, circle, sphere or cylinder is quite common, and we use linear algebra, matrix methods, and non-linear optimization techniques for this purpose. Fitting is therefore one "processing" technique in engineering metrology data analysis. It should be pointed out that fitting is a critical analysis step in coordinate metrology, where features of interest such as size, parallelism, perpendicularity *etc.* are determined from substitute geometry; fitting is used primarily as a preprocessing step in surface texture analysis.

Subsequent to any preprocessing by fitting, the data undergoes an information extraction phase, where suitable surface features are extracted because of their relevance to functional properties of the component. This step is referred to as *filtering*. The underlying assumption here is that engineering surfaces are comprised of surface wavelengths, and that certain wavelength bandwidths have specific bearings on certain aspects of a component's functionality. Therefore, filtering involves partitioning a profile or topography into different wavelength bandwidths. The fundamental basis for filtering lies in signal-processing theory.

Parameters are then computed on fitted and filtered surface and roundness data. These parameters are toleranced during design, and as mentioned before, they are inspected during quality control. A variety of parameters exists, some trivial to compute and others not. Therefore *parameterization* is a third topic of computational relevance.

Finally, in numerical computations, *uncertainty* in the inputs propagates through the chain of computations, and is impacted by the algorithms used, and the type and nature of the computations themselves. This is an important area in metrology and has its bearing in probability and statistics. Uncertainty analysis is generally not provided by commercial instrumentation, but national measurement labs do provide such analysis with their reference software.

In summary, data processing generally falls into the four categories discussed above, namely, fitting, filtering, parameterization, and uncertainty (Fig. 1.1), and this book exclusively addresses these topics. Some of the material presented here is available in the literature in journals and conference proceedings, some as early as 1944, and some as recent as 2007. Many of the details are buried in personal communications that researchers have had with colleagues over the last several decades. Some mathematics may be discussed in other textbooks, and in national and international Standards, but in a manner that is sometimes abstract and unsuitable for immediate implementation.

After more than six decades of the practice of surface and roundness metrology, there is still no single detailed compilation of the mathematical techniques used. We felt the need for such a book during our research and teaching activities and our effort here hopefully addresses some of that need. In each of the chapters, we describe the mathematics and illustrate the concepts through worked examples. We also provide end-of-chapter exercises with answers at the end of the text. We use MATLAB (a product of The MathWorks) extensively through out the text. Several excellent tutorials are available for those unfamiliar with MATLAB; the MathWorks website is a good reference (MATLAB).

1.3 Organization

We have organized the remaining chapters into five parts. Although fitting precedes filtering in the analysis hierarchy, we discuss filtering first (Part I) because of its significance to the ultimate goal of surface texture analysis, namely, the correlation of surface texture parameters and functional behavior of components. In Chaps. 2, 3, and 4, we present a historical context to filtering and then discuss implementation aspects from both a frequency- and time-domain perspective. Chapters 5 and 6 are devoted to the Gaussian and the two-resistor-capacitor (2RC) filters. In each case, we illustrate filter characteristics through discussions and worked examples. Then, we extend the filtering ideas to two cases: roundness and 3D data analysis in Chaps. 7 and 8 respectively.

In Part II, we discuss some recent advances in filtering techniques, such as the Gaussian regression filter, the spline filter, robust filters, morphological filters, and multi-scale filters. Each has its own unique advantages and disadvantages; we illustrate these with worked examples using simulated surface profile data.

Part III is devoted to the next computational topic of importance, namely, fitting. Fitting reference circles is a critical aspect of roundness analysis. Fitting lines,

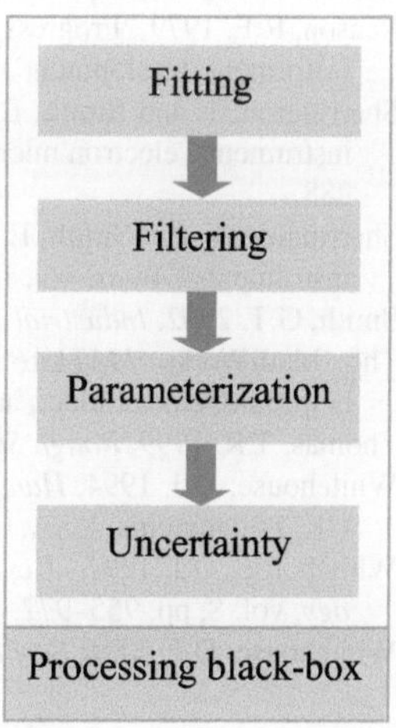

Fig. 1.1 Data processing in surface and roundness metrology

planes, circles, spheres, and cylinders may sometimes be used in preprocessing surface texture data as well. We discuss these topics in this part. After a brief introductory chapter (Chap. 14), we elaborate procedures for fitting least-squares lines and planes (Chap. 15). Non-linear methods that are useful for circles, spheres, and cylinders are presented in Chaps. 16 and 17. Then, a linear approximation for circles useful for roundness analysis is illustrated (Chap. 18), followed by a discussion on exchange algorithms for minimum-zone lines, planes, and circles (Chap. 19). Finally, we wrap up Part III with Chap. 20, a discussion of linear-programming-based simplex for circle fitting and illustrate how this method is linked to exchange algorithms discussed in the previous chapter, Chap. 19.

The focus of Part IV is on parameters that can be computed from surface profiles. These include amplitude, spatial, and hybrid parameters. We also illustrate the computation of statistical functions such as the autocorrelation function, the power spectral density, and the bearing area curve. The final chapter, Chap. 23 in this part is devoted to parameters and statistical functions computed from 3D topography.

In Part V, we discuss topics concerning error and uncertainty. We first discuss concepts related to uncertainty propagation and its implications in surface profile filtering and surface fitting (Chaps. 24 and 25). In Chap. 26, we discuss methods for separating spindle error from part error in roundness analysis. Finally, in the concluding Chap. 27, we take a look at some emerging computational topics that are finding relevance in surface and roundness metrology.

References

Jiang, X., Scott, P.J., Whitehouse, D.J. and Blunt, L. 2007, 'Paradigm shifts in surface metrology. Part I. Historical philosophy', *Proceedings of the Royal Society A Mathematical Physical and Engineering Sciences*, vol. 463, no. 2085, pp. 2049–2070.

Reason, R.E., Hopkins M.R. and Garrod R.I. 1944, *Report on the Measurement of Surface Finish by Stylus Methods*, Rank Organization. Leicester, UK.

Reason, R.E. 1966, *Report on the Measurement of Roundness*, Rank Organization. Leicester, UK.

Reason, R.E. 1979, 'Progress in the appraisal of surface topography during the first half-century of instrument development', *Wear*, vol. 57, pp. 1–16.

Sherrington, I. and Smith, E.H. 1988a, 'Modern techniques in surface metrology: part I; stylus instruments, electron microscopy and non-optical comparators', *Wear*, vol. 125, no. 3, pp. 271–288.

Sherrington, I. and Smith, E.H. 1988b, 'Modern techniques in surface metrology: part II; optical instruments', *Wear*, vol. 125, no. 3, pp. 289–308.

Smith, G.T. 2002, *Industrial Metrology: Surfaces and Roundness*, 1st edn, Springer. London, UK.

The MathWorks *MATLAB documentation*, available at http://www.mathworks.com/access/helpdesk/help/techdoc/, accessed Jan 25 2008.

Thomas, T.R. 1999, *Rough Surfaces*, 2nd edn, Imperial College Press. London, UK.

Whitehouse, D.J. 1994, *Handbook of Surface Metrology*, Institute of Physics Publishing. Bristol, UK, Philadelphia, USA.

Whitehouse, D.J. 1997, 'Review article: Surface Metrology', *Measurement Science and Technology*, vol. 8, pp. 955–972.

Whitehouse, D.J. 2004, *Surfaces and Their Measurement*, Kogan Page Science. London, UK.

Part I
Filtering

We discuss filtering techniques as applied to 2D surface profiles, roundness profiles, and 3D surface topography in this part. We trace the evolution of different filtering methods in Chap. 2. In the Chaps. 3 and 4, we introduce basic ideas involved in frequency- and time-domain filtering. In Chaps. 5 and 6, we focus our attention on two commonly used filters: the Gaussian and the 2RC filter. In the final two chapters in this part, Chaps. 7 and 8, we describe filtering as applied to roundness profiles and 3D surface topography.

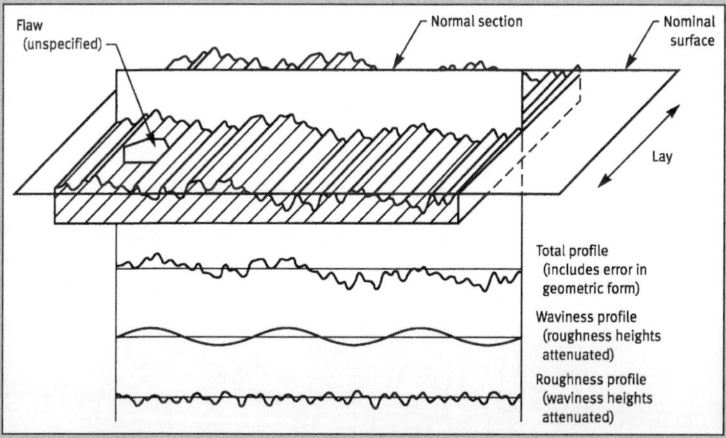

Fig. I Surface roughness and waviness. Reprinted from ASME B46.1-2002, by permission of the American Society for Mechanical Engineers. All rights reserved

Chapter 2
A Brief History of Filtering

2.1 Introduction

Engineering surfaces are viewed as having fine texture called roughness, superimposed on more general curvature called waviness, and long-range deviations called form. This has been recognized from the early days of surface texture measurement. The motivation for such classification comes from the fact that roughness, waviness, and form have different origins and affect part functionality in different ways (Whitehouse 1994). Therefore, partitioning a profile into roughness, waviness, and form is of primary significance and an important aspect of surface texture analysis. This partitioning is achieved through a process called filtering.

In the infancy of surface texture measurements, profile data was filtered graphically (Whitehouse and Reason 1965). The raw profile was divided into segments of equal length, and in each segment, a mean line was drawn that captured the slope of the profile in that segment. The roughness profile was obtained by considering the deviations of the points from these mean lines. The graphical approach was cumbersome and time-consuming. An automated way to deriving the mean line became necessary and electrical filters were soon adopted for surface texture filtering.

2.2 Electrical Filters

In electrical filtering, a voltage proportional to the profile is passed through a two-resistor-capacitor (2RC) network. Because the 2RC network has memory, the output is not only a function of the input at any instant of time, but also a function of prior values. In effect, the 2RC network computes a running average of current and previous voltages, but gives reduced weights to voltages more distant in the past. While the memory of the network helps in averaging, by remembering past values only, it introduces an undesired phase in the output.

2.3 Digital Filters

To understand this phase behavior and to rectify the problem, Whitehouse and Reason (1965) simulated the 2RC filter digitally. They described the filter using a weighting function that depended on the cutoff. A cutoff serves a similar purpose as the size of a segment in graphical analysis, namely to separate the short-wavelength components from the long-wavelength components.

They were able to show that the weighting function can be convolved with the raw profile to produce a running average mean line. The digital mean line thus obtained was identical to the hardware electrical filter mean line. The digital implementation of the 2RC filter was a significant step forward. With increasing computing capabilities, the electrical filter quickly became obsolete and filtering began to be accomplished using computers. Raja and Radhakrishnan (1977, 1979a, 1979b) have explored digital filtering technique extensively. They have used the Fourier transform for filtering surface profiles and also demonstrated filtering in the time domain using impulse response (the *finite impulse response* of a filter), in several early papers noted previously.

With the advent of digital filtering, researchers also began to address the primary shortcoming of the 2RC filter, namely its non-linear phase. A phase-correct 2RC filter was designed (Whitehouse and Reason 1965; Whitehouse 1967). Subsequently, the Gaussian filter was introduced, and it remains the preferred choice of filter to this day. Vorburger and Raja (1990) discuss some of these developments in their report.

2.4 The Envelope Method

Around the time that researchers at Rank Taylor Hobson were developing the 2RC filter, Von Weingraber (1956, 1957) in Germany was working on an alternative technique to separate the fine texture from the overall form. In many applications, surface texture plays an important role during contact between mating surfaces. Therefore, on a measured profile, he simulated this contact by determining the envelope of a ball rolling on a surface. The deviations from the envelope would be the fine texture or roughness.

This system came to be known as the E system, while the electrical filter-based technique came to be termed the M system. Over the years, the M system became the *de-facto* filtering choice. More recently, the E system has re-emerged and found prominence as part of a larger group of techniques referred to as morphological filters.

2.5 The Gaussian Filter

The next major development in surface texture filtering can be attributed to the arrival of the Gaussian filter (ISO 11562:1996, ASME B46.1-2002). It solved the problem of non-linear phase of the 2RC and could be implemented digitally quite easily.

2.6 Overlap of Measurement Techniques

Today, with the availability of increasingly sophisticated gages, sensors, and powerful data-processing capabilities using personal computers, there is an overlap of measuring capability. A typical stylus-based instrument can capture roughness, waviness, and form. A roundness-measuring instrument can also gather straightness data and a coordinate measuring machine (CMM) can acquire both dimensional information and form. Figure 2.1 shows the current metrology measurement spectrum. As the bandwidth of measurement instruments increases, it becomes essential to separate surface profile data into meaningful wavelength regimes before numerical characterization.

2.7 Recent Advances in Filtering

In the last decade, there have been some significant advances in filtering techniques; reviews can be found in Raja *et al.* (2002) and Muralikrishnan (2003). The Gaussian filter, which is widely used today, suffers from several shortcomings. First, the waviness profile has distortion near the edges (this is true for open profiles, not for roundness). In fact, a length equal to half a cutoff or one cutoff is typically removed from the beginning and from the end of both the roughness and waviness profiles to limit the effect of edge distortion. For already-short traces, discarding an additional one cutoff at both ends is not an acceptable solution.

Another issue with the Gaussian filter is that the waviness profile does not follow the texture in the presence of large form, such as when measuring on nominally

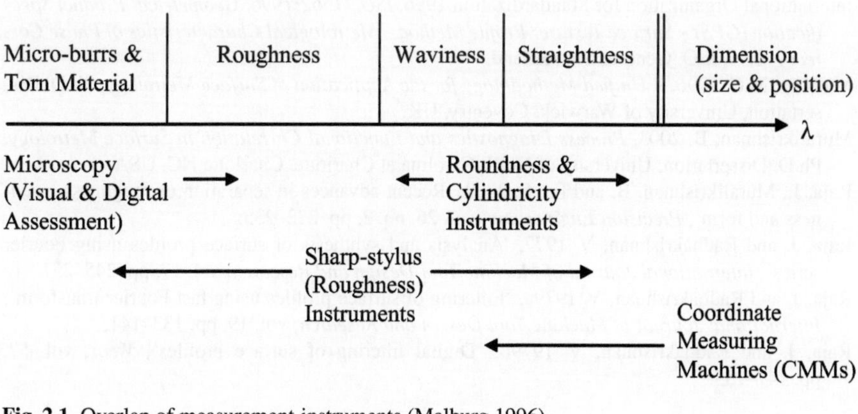

Fig. 2.1 Overlap of measurement instruments (Malburg 1996)

cylindrical or spherical geometry. It is quite common to first remove the best-fit nominal surface before applying the filter for this purpose. A third drawback of the Gaussian filter is that it is not robust against outliers. A large peak or valley will distort the waviness profile in the vicinity of an outlier.

The spline and Gaussian regression filters overcome the problems of edge distortion and poor performance in the presence of large form, while the robust spline and robust Gaussian regression filters overcome all three problems. Wavelet-based filters are also finding increasing applications in surface texture measurement because they provide an efficient way to partition a profile into multiple- and narrow-wavelength bands, something not feasible with the Gaussian filter. And as mentioned earlier, morphological filters and scale-space techniques provide an entirely different perspective to filtering, and are finding new and exciting applications. We will discuss these techniques in later chapters.

2.8 Summary

We have briefly traced the evolution of filtering techniques in surface metrology. In this and subsequent parts, we will discuss many of these topics in considerable detail. In the next few chapters, we discuss some elementary concepts in filtering with the objective of allowing the reader to comprehend the mathematics and perform implementations of commonly used filters such as the Gaussian or the 2RC. In Part II, we discuss more recent advances such as the Gaussian regression, spline, morphological, and multi-scale filters.

References

American Society of Mechanical Engineers 2002, *ASME B46.1-2002, Surface Texture (Surface Roughness, Waviness and Lay)*, ASME. New York, NY, USA.

International Organization for Standardization 1996, *ISO 11562:1996, Geometrical Product Specification (GPS) – Surface Texture: Profile Method – Metrological Characteristics of Phase Correct Filters*, ISO. Geneva, Switzerland.

Malburg, M.C. 1996, *A Unified Methodology for the Application of Surface Metrology*, Ph.D. Dissertation, University of Warwick. Coventry, UK.

Muralikrishnan, B. 2003, *Process Diagnostics and Functional Correlation in Surface Metrology*, Ph.D. Dissertation, University of North Carolina at Charlotte. Charlotte NC, USA.

Raja, J., Muralikrishnan. B. and Fu. S. 2002, 'Recent advances in separation of roughness, waviness and form', *Precision Engineering*, vol. 26, no. 2, pp. 222–235.

Raja, J. and Radhakrishnan, V. 1977, 'Analysis and synthesis of surface profiles using Fourier series', *International Journal of Machine Tool Design and Research*, vol. 17, pp. 245–251.

Raja, J. and Radhakrishnan, V. 1979a, 'Filtering of surface profiles using fast Fourier transform', *International Journal of Machine Tool Design and Research*, vol. 19, pp. 133–141.

Raja, J. and Radhakrishnan, V. 1979b, 'Digital filtering of surface profiles', *Wear*, vol. 57, pp. 147–155.

Von Weingraber, H. 1956, 'Zur Definition der Oberflächenrauheit Werkstattstechnik', *Masch. Bau*, vol. 46.

Von Weingraber, H. 1957, 'Über die Eignung des Hullprofils als Bezugslinie für Messung der Rauheit', *Microtechnic*, vol. 11, pp. 6–17.

Vorburger, T.V. and Raja, J. 1990, *Surface Finish Metrology Tutorial*, NISTIR 89-4088, National Institute of Standards and Technology. Gaithersburg, MD.

Whitehouse, D.J. 1967, 'Improved type of wavefilter for use in surface-finish measurement', *Proceedings of the Institution of Mechanical Engineers*, vol. 182, no. 3K, pp. 306–318.

Whitehouse, D.J. 1994, *Handbook of Surface Metrology*, Institute of Physics Publishing. Bristol, UK, Philadelphia, USA.

Whitehouse, D.J. and Reason, R.E. 1965, *The Equation of the Mean Line of Surface Texture Found by an Electric Wave Filter*, Rank Taylor Hobson. Leicester, UK.

Chapter 3
Filtering in the Frequency Domain

3.1 Surface Wavelengths

Every engineering surface can be viewed as comprising of sinusoidal functions of different amplitudes and wavelengths. In fact, roughness is simply a collection of all high-frequency or small-wavelength sinusoidal components of a surface, waviness is a collection of all medium-wavelength components, and form consists of all large-wavelength components.

The transition from roughness to waviness, or from waviness to form is determined by user-defined wavelength cutoffs. Because roughness, waviness, and form represent certain defined wavelength bandwidths, two cutoffs are required to properly establish each band, one for the lower bound and another for the upper bound. Thus, cutoffs λcw (waviness long-wavelength cutoff) and λsw (waviness short-wavelength cutoff) (ASME B46.1-2002) define waviness while λc (roughness long-wavelength cutoff) and λs (roughness short-wavelength cutoff) define roughness. The values chosen for the cutoffs depend on the application; the ASME B46.1 Standard (American Society of Mechanical Engineers 2002) provides some guidelines for selection of cutoff ratios, stylus tip sizes, and traverse lengths.

In practice, however, the measurement instrument capability or measurement setup itself applies a filter that establishes one of these cutoffs. Therefore only one cutoff is specified to define roughness or waviness. For example, the finite stylus size may establish the roughness short-wavelength cutoff through mechanical filtering. Therefore, a roughness long-wavelength cutoff applied to the measured profile may be sufficient to obtain roughness.

A surface profile, like most other digitized signals of observable physical quantities (voltage, current etc), exists in the time or spatial domain, because the signal is monitored as a function of time (plot of voltage as a function of time) or spatial location (plot of surface profile as a function of spatial location). From Fourier theory, we know that such signals can be viewed as comprising of sinusoids of varying frequencies, amplitudes, and phases. In other words, by transforming a surface profile signal to the frequency domain using the Fourier transform, we can clearly identify the amplitude and phase of sinusoids of specific wavelengths.

This transformation is very useful, because by viewing the information contained in the profile, not as a function of spatial location, but as a function of sinusoidal wavelengths, we can easily suppress or retain selected wavelength ranges, before transforming back to the spatial/time domain. This process of transforming a profile to the frequency domain, suppressing certain frequency ranges, and transforming back to the spatial domain is the basic idea of the filtering process. Before we discuss details, let us briefly review Fourier transform and its implementation.

3.2 Fourier Transform, Discrete Fourier Transform, and Fast Fourier Transform

The Fourier transform $\tilde{z}f$ of a continuous and infinitely long signal $\tilde{z}$ is given by the integral

$$\tilde{z}f(\zeta) = \int_{-\infty}^{\infty} \tilde{z}(x)e^{-i2\pi x\zeta}\mathrm{d}x\,, \tag{3.1}$$

where ζ is the frequency and $i = \sqrt{-1}$. Because surface profiles are digitized finite duration signals, we apply the discrete Fourier transform (DFT), which is given by

$$zf(f) = \sum_{j=1}^{n} z(j)e^{-i2\pi f(j-1)\Delta x}\,, \tag{3.2}$$

where z is the digitized signal containing n points, Δx is the sampling interval, and zf is the DFT of z. The spatial frequency, f, is equal to $(k-1)/l$, where k is an integer ranging from 1 to $n/2+1$ if n is even, and $(n-1)/2+1$ if n is odd, and l is the profile length and is equal to $n\Delta x$. The surface wavelength λ is equal to $1/f$.

Attention to terminology will help avoid confusion. The Fourier transform is associated with continuous functions. The discrete Fourier transform is associated with finite duration digitized signals and is given by Eq. 3.2. It can be implemented in several ways. The direct implementation is to compute the sum of the products for each frequency term f. This is an inefficient method. The fast Fourier transform (FFT) is an algorithm that efficiently implements the DFT (Press *et al.*, 1992). The discrete-time Fourier transform (DTFT) is a limiting case of the DFT when the length is allowed to go to infinity (Oppenheim *et al.*, 1999). We will illustrate the computation of the DFT using the following examples.

Example 3.1. Generate two sinusoidal profiles z_1 (amplitude 1 μm, wavelength 2 mm) and z_2 (amplitude 0.5 μm, wavelength 0.5 mm) defined in the interval $0 \leqslant x < 8$ mm and digitized using a stylus instrument with a sampling interval $\Delta x = 1$ μm. Then generate profile z as the sum of z_1 and z_2. We will use profile z and other such simulated surface data in examples and exercises.

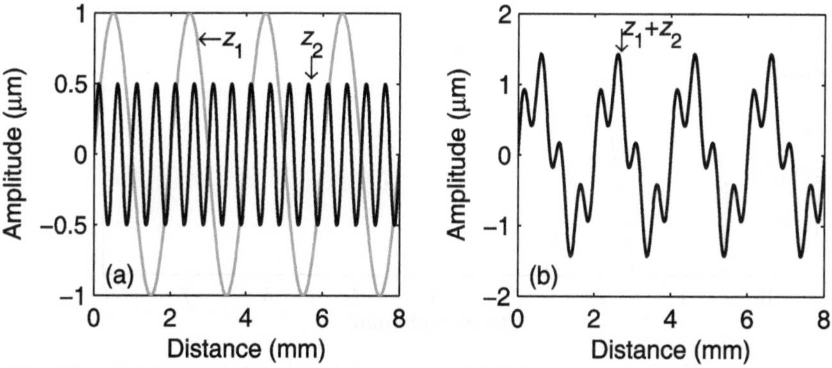

Fig. 3.1 **a** Profiles z_1 and z_2. **b** Profile $z = z_1 + z_2$

Profile z is generated as shown below. Figure 3.1a shows profiles z_1 and z_2 and Fig. 3.1b shows profile z.

```
dx=0.001;                   % spacing in mm
length=8;                   % profile length in mm
x=(0:dx:length-dx)';
z1=1*sin(2*pi.*x/2);        % amplitude 1 µm, wavelength 2 mm
z2=0.5*sin(2*pi.*x/0.5); % amplitude 0.5 µm, wavelength 0.5 mm
z=z1 + z2;                  % generate z
figure(1); plot(x,z1,'r',x,z2,'b');
xlabel('Distance (mm)');
ylabel('Amplitude (\mum)')
figure(2); plot(x,z,'k');
xlabel('Distance (mm)');
ylabel('Amplitude (\mum)');
```

Example 3.2. Plot the DFT of profile z

The discrete Fourier transform zf of profile z can be obtained by calling MATLAB's `fft` function. Care must be taken in interpreting the meaning of the different elements of zf (Press *et al.*, 1992). Each element of zf is complex, and the magnitude and phase of each complex number is itself the amplitude and phase of a sinusoid of specific wavelength. The wavelength, however, has to be interpreted from the position of the element in the array. The first element of zf is the constant term (corresponding to a frequency of zero, or DC). The jth element corresponds to a wavelength of $l/(j-1)$, where j ranges from 2 to $(n/2+1)$ if n is even, or from 2 to $[(n-1)/2+1]$ if n is odd (n is the length of array z). The second half of array zf is a mirror of the first half. Therefore, arrays zf and z have equal lengths. The first half of zf contains elements that represent decreasing wavelengths. The largest wavelength measurable is the measurement length l, and the smallest wavelength measurable is twice the sampling interval (Nyquist frequency). The code segment below interprets and plots the DFT of profile z (see Fig. 3.2). Notice that the amplitude of zf is scaled by $n/2$. This follows directly from Eq. 3.2.

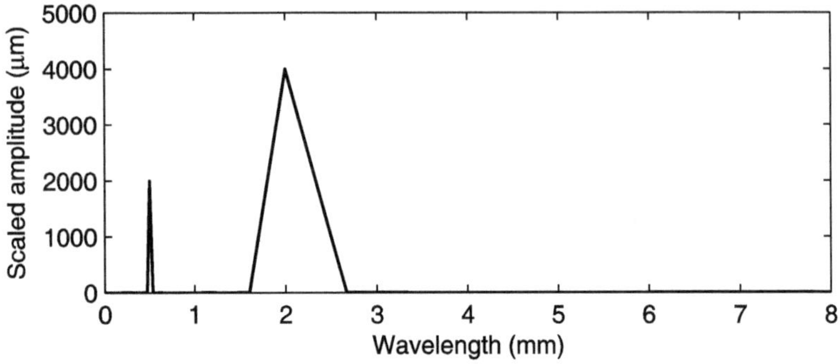

Fig. 3.2 Discrete Fourier transform of profile z (the y axis is $n/2$ times the amplitude of the input sinusoid)

```
dx=0.001;                 % spacing
l=8;                      % length of the profile
n=l/dx;
zf=fft(z);                % perform FFT
j=(2:1:floor(n/2)+1)'; % generate the wavelength array
lambda=n*dx./(j-1);       % generate the wavelength array
plot(lambda,abs(zf(2:floor(n/2)+1,1))); % plot half of the
                                        % FFT array
xlabel('Wavelength (mm)');
ylabel('Scaled amplitude (\mum)');
```

Having plotted the DFT, we next verify the amplitude and phase characteristics of the individual components z_1 and z_2. The 5th term of zf represents a wavelength of 2 mm and the 17th term represents a wavelength of 0.5 mm. Therefore, the amplitude and phase of these terms can be obtained as shown below. Notice that although the phase is not zero, it is identical for both sinusoids.

```
Amplitude_z1=abs(zf(5,1))/(n/2)   % amplitude of z1
Phase_z1=angle(zf(5,1))           % phase of z1
Amplitude_z2=abs(zf(17,1))/(n/2) % amplitude of z2
Phase_z2=angle(zf(17,1))          % phase of z2
```

3.3 Filtering in the Frequency Domain

We have seen how to transform a surface profile from the spatial domain to the frequency domain. Having determined the frequency content of the input signal, we define a filter whose purpose is to suppress all wavelengths smaller than an arbitrarily chosen cutoff of 0.8 mm. This particular value is chosen instead of say, 1 mm, because it not only achieves our filtering objective but is also a recommended value from the ASME B46.1 Standard (American Society of Mechanical Engineers 2002). This filter is shown in the DFT plot in Fig. 3.3 as an array of zeros and

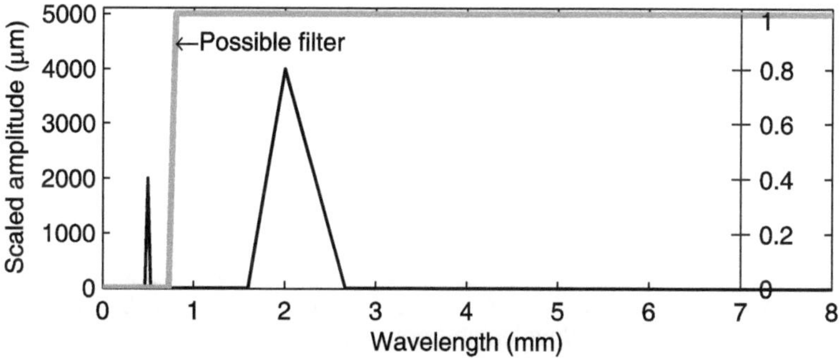

Fig. 3.3 A possible filter defined in the frequency domain (filter scale plotted on the *right*)

ones (the scale for the filter is shown on the right), with the transition from 0 to 1 occurring at a wavelength of 0.8 mm.

The purpose of the filter is to define weights that attenuate the amplitude of sinusoids of different wavelengths. The filter curve shown in Fig. 3.3 is therefore called the amplitude transmission characteristics of the filter. In order to define a filter completely in the frequency domain, the phase offset of different sinusoidal wavelengths has to be specified also. The combination of amplitude and phase characteristics of a filter is referred to as the transmission characteristics of the filter.

After the filter is defined, the DFT of profile z is multiplied with the transmission characteristics of the filter. The inverse transform then provides the filtered profile. The following examples illustrate the frequency-domain approach to filtering profiles.

Example 3.3. Generate a filter Sf with step transmission characteristic in frequency domain with a cutoff of 0.8 mm.

The objective is to build an array Sf of zeros and ones, with the implicit understanding that each element in the array corresponds to the transmission of a specific wavelength. Ultimately we intend to multiply Sf with the DFT of the profile (zf) and therefore, the elements of Sf must be arranged in the same order of wavelengths as that of zf. Therefore, the first element of Sf must be the DC term, which can be set to 0. The second element of Sf must correspond to a wavelength equal to the length of the profile (8 mm in this case). Because this wavelength is larger than 0.8 mm, the filter will transmit this wavelength and therefore the value of this element is 1. The third element of Sf corresponds to a wavelength of half the length of the profile (4 mm). The value of this element is also 1. This process is repeated and appropriate values for each element are decided based on the wavelength that particular element represents. If the cutoff is 0.8 mm, elements whose indices range from 2 to 11 have values of 1. Also, recall that the wavelengths decrease to the middle and increase again. Therefore, the elements whose indices range from 7,991 to 8,000 also have a value of 1. All other elements

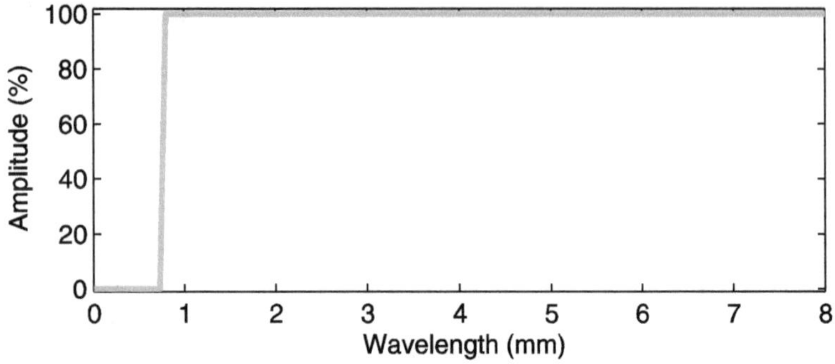

Fig. 3.4 Amplitude transmission characteristics of filter Sf

have a value of 0. The filter array Sf is plotted in Fig. 3.4 (only one half of Sf is shown).

```
j=(2:1:floor(n/2)+1)';
lambda=n*dx./(j-1);           % generate wavelength axis
k=find(lambda >= 0.8);        % find indices of elements that
                              % represent wavelength >= 0.8 mm
Sf=zeros(size(lambda,1),1);   % create an array of zeros
Sf(k)=1;                      % assign a value of 1 for those
                              % terms that correspond to
                              % wavelength >= 0.8 mm
% construct the filter array for the second half (mirror terms)
if mod(n,2) == 0
Sf=[0; Sf; flipud(Sf(1:size(Sf,1)-1))];
else
Sf=[0; Sf; flipud(Sf)];
end
% plot the amplitude as a percentage
plot(lambda,100*abs(Sf(2:floor(n/2)+1,1)));
axis([0 8 -10 110])
xlabel('Wavelength (mm)');
ylabel('Amplitude (%)');
```

While we have explicitly described the amplitude transmission of the different wavelength components, we have also implicitly controlled the phase offset. This is because each element is a complex number whose amplitude represents the amplitude attenuation of a sinusoid of specific wavelength, and whose angle represents the phase offset. Therefore a value of 1 represents an amplitude attenuation of unity and a phase of 0, while a value of 0 represents an amplitude attenuation of 0 and a phase of 0.

Example 3.4. Apply filter Sf on zf and generate the filtered profile.

The filtering process now involves a term by term multiplication of zf with Sf, followed by an inverse transform to obtain the spatial-domain signal. The following MATLAB segment illustrates this. The filtered profile w is plotted along with the

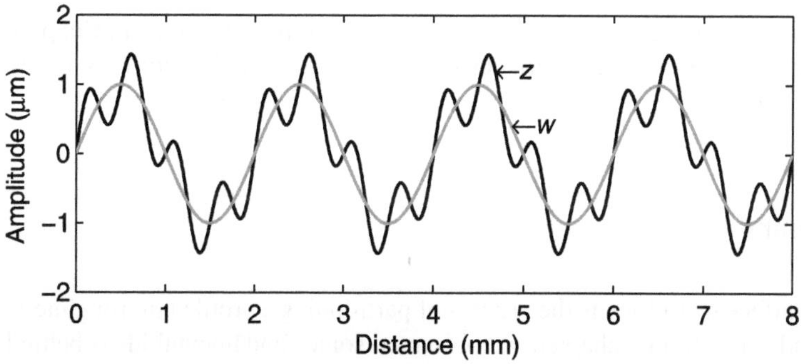

Fig. 3.5 Input profile z and filtered profile w

input profile z in Fig. 3.5. Notice that the filtered profile contains only the larger wavelength component of z, as expected.

```
temp=zf.*Sf;           % term by term multiplication
w=real(ifft(temp));   % extract the real part of ifft
plot(x,z,'r',x,w,'b');
xlabel('Distance (mm)');
ylabel('Amplitude (\mum)');
```

3.4 Wrap-Around Effect

Although we have successfully filtered a profile in the preceding examples, there is an important point to note, one which we will revisit in later chapters. When performing DFT, the profile is considered to be periodic; that is, it wraps around itself as if it were a closed profile. For roundness profiles, this is beneficial, but for open profiles such as surface profiles on a flat surface, this poses a problem. A common solution is to add sufficient zeros to the beginning and end of the profile, so that even when the profile does wrap around, the filtering process is immune to this effect.

3.5 Amplitude Transmission and Phase Characteristics

In the example above, we defined a filter with step transmission. That is, the filter transmitted all wavelengths above a certain cutoff without attenuating its amplitude. Such a step filter, while highly desirable, is rarely used in surface metrology because it has no time-domain representation. A functional representation in the time domain is useful as we will see in Chap. 4. Instead, filters with a more gradual transmission are generally used.

A filter can not only attenuate the amplitude of an input sinusoid, it can also affect its phase. A filter that produces different phase offsets for different input wavelengths is not desirable because the filtered output will have distortions. In Chap. 6, we will look at the phase characteristics of the 2RC filter and distortions in the roughness profile.

3.6 Summary

Filtering in surface metrology is the process of partitioning a profile into roughness, waviness, and form. In this chapter, we presented some fundamental ideas behind filtering. We transformed the profile into the frequency domain so that the information contained in the profile can be viewed as a function of surface wavelengths instead of as a function of spatial location. We defined a filter in the frequency domain by specifying its amplitude and phase transmission characteristics. The filter was applied on the transformed profile and an inverse transform yielded the filtered profile in the time domain. Without explicitly detailing the underlying theory, we have discussed several concepts from signal processing. We encourage the reader to review early chapters in Oppenheim *et al.* (1999) and also Ingle and Proakis (1996) and McClellan *et al.* (1997). In Chap. 4, we will review an alternative and more convenient approach to filtering, and show how the methods described there and in this chapter are related.

Exercises

3.1. Compute the DFT of array $z = [1\ 6\ 3\ 2]$ (in units of micrometers) by writing a simple program that evaluates Eq. 3.2 explicitly. Assume that z contains the vertical heights of a surface profile at $x = [1\ 2\ 3\ 4]$ (x in units of millimeters). Verify the result using the `fft` function in MATLAB.

3.2. Generate a profile z containing two sinusoidal components: z_1 of amplitude 0.5 μm, wavelength 2 mm, zero phase, and z_2 of amplitude 1 μm, wavelength 0.5 mm and 0.5 rad phase. Let spacing $\Delta x = 1$ μm and profile length $l = 8$ mm. Compute the DFT of profile z and verify the amplitude and relative phase of the individual components.

3.3. Generate and plot the amplitude and phase characteristics of a filter whose transmission characteristics is given by:

$$Sf_1 = e^{\left(-\pi\left(\alpha\frac{\lambda c}{\lambda}\right)^2\right)}, \quad 0 \leqslant \lambda < 8\,\text{mm}\,, \quad \alpha = 0.4697$$

Let $\lambda c = 0.8$ mm. Does the filter have linear or non-linear phase? What is the amplitude attenuation at the cutoff? (The transmission characteristics belong to a low-pass Gaussian filter; we define low-pass and high-pass filters in Chap. 5).

3.4. Generate and plot the amplitude and phase characteristics of a filter whose transmission characteristics is given by:

$$Sf_2 = \frac{1}{\left(1 + \frac{j\lambda c}{\lambda\sqrt{3}}\right)^2}, \quad 0 \leqslant \lambda < 8\,\text{mm}$$

Let $\lambda c = 0.8$ mm. Does the filter have linear or non-linear phase? What is the amplitude attenuation at the cutoff? (The transmission characteristics belong to a low-pass 2RC filter).

3.5. Generate and plot the amplitude and phase characteristics of a filter whose transmission characteristics is given by:

$$Sf_3 = \frac{1}{\left(1 - \frac{j\lambda}{\lambda c\sqrt{3}}\right)^2}, \quad 0 \leqslant \lambda < 8\,\text{mm}$$

Let $\lambda c = 0.8$ mm. Does the filter have linear or non-linear phase? What is the amplitude attenuation at the cutoff? (The transmission characteristics belong to a high-pass 2RC filter).

References

American Society of Mechanical Engineers 2002, *ASME B46.1-2002, Surface Texture (Surface Roughness, Waviness and Lay)*, ASME. New York, NY, USA.

Ingle, V.K. and Proakis, J.G. 1996, *Digital Signal Processing Using MATLAB V.4*, PWS Publishing Company. Boston, MA, USA.

McClellan, J.H., Burrus, C.S., Oppenheim, A.V., Parks, T.W., Schafer, R.W. and Schuessler, H.W. 1997, *Computer-Based Exercises for Signal Processing Using MATLAB 5*, Prentice Hall. Upper Saddle River, NJ, USA.

Oppenheim, A.V., Schafer, R.W. and Buck, J.R. 1999, *Discrete-Time Signal Processing*, 2nd edn, Prentice Hall. Upper Saddle River, NJ, USA.

Press, W.H., Flannery, B.P., Teukolsky, T.A. and Vetterling, W.T. 1992, *Numerical Recipes in C: The Art of Scientific Computing*, 2nd edn, Cambridge University Press. Cambridge, UK.

Chapter 4
Filtering in the Time Domain

4.1 Filtering as an Averaging Process

In Chap. 3, we discussed a frequency-domain approach to filtering. That method was intuitive, but somewhat cumbersome to implement. In this chapter, we will discuss a time-domain approach to filtering and show that the two methods are closely related.

Filtering can be viewed as an averaging process applied to the noisy original surface profile to obtain a smoother profile. This process of averaging is analogous to moving window averages used in monitoring stock market fluctuations. In the stock market example, if the price history of a stock is viewed over the last five years without any averaging, the chart is noisy and long term trends are not discernible. On the other hand, if the price is averaged over a two week or one month window, the long term trend is evident. In the same manner, the raw surface profile is noisy and does not reveal the underlying medium- and long-wavelength components clearly. A moving window average applied to the profile extracts the long-wavelength components by smoothing short-term fluctuations.

We will first discuss implementation details of this moving average method and then illustrate how this method is tied to the approach discussed in Chap. 3. Consider the following simple example in Fig. 4.1 that explains this moving average computation. A profile z containing six sampling elements is smoothed using a filter S (also called the weighting function) that averages three elements at a time.

The process begins by positioning the weighting function at the left end of the profile so that the profile just overlaps the weighting function. A weighted average is then computed to obtain the first term of the filtered profile. Then, the weighting function is stepped over to the right by one element and the weighted average is evaluated again to obtain the second element of the filtered profile. These steps are repeated until the weighting function is positioned at the extreme right, and just overlaps the profile.

There are several interesting points to note. First, the filtered output is longer than the original profile. In fact, the length of the filtered profile is one less than

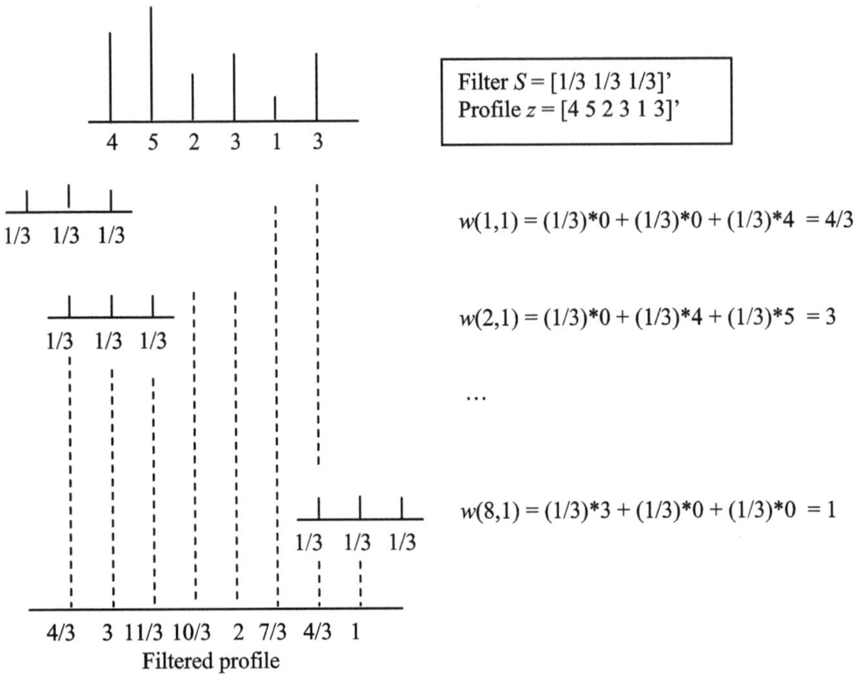

Fig. 4.1 Moving average computations to obtain filtered profile

the sum of the lengths of the original profile and the weighting function. Therefore, half the length of the weighting function (rounded down to the nearest integer) can be discarded from the beginning and from the end of the filtered profile. Second, the length of the weighting function is an important parameter in the filtering process. Typically, the length of the weighting function is a function of a variable called cutoff. As the cutoff increases, the weighting function size increases and therefore the resulting output is smoother. Third, notice that the weights are all identical in this example and that the sum of weights is unity. This weighting function represents a rectangular filter. A triangular filter provides the greatest weight to the central element with the weights reducing linearly from the center to the edges. A Gaussian filter also provides the greatest weight to the central element, but the weights taper off gradually. As a final point to note, the process of averaging always results in a smoother profile. The high-frequency portion of the profile can be obtained by subtracting the result of averaging from the input profile.

The moving average discussed here is also referred to as convolution in signal processing. MATLAB provides an inbuilt function `conv` that performs this moving average computation (a strict definition of convolution involves mirroring the weighting function prior to computing the moving average. If the weighting function is symmetric, mirroring is unnecessary; see Oppenheim *et al.* (1999) for details).

Example 4.1. Verify convolution computations in Fig. 4.1 using MATLAB.

Define arrays z and S, and call MATLAB's `conv` function as shown below.

```
z=[4 5 2 3 1 3]';
S=[1/3 1/3 1/3]';
conv(z,S)
```

Example 4.2. Generate a rectangular weighting function S of width 0.8 mm, sampled at spacing $\Delta x = 1$ μm, whose elements have a value of 1. Normalize the weighting function so that the sum of all weights is unity.

Weighting function S is generated as shown below. Notice that the weighting function is plotted in the interval -0.4 mm $\leqslant x \leqslant 0.4$ mm. This is a matter of convention. The weighting function may also be plotted in the interval $0 \leqslant x \leqslant 0.8$ mm. The result of convolution is not affected by this.

```
alpha=0.4;
dx=0.001;                % spacing
x=(-alpha:dx:alpha)'; % generate the x axis
S=ones(size(x,1),1);  % generate the weighting function
S=S/sum(S);           % weights are normalized to unit sum
plot(x, S)
```

Example 4.3. Convolve profile z from Example 3.1 (in Chap. 3), with weighting function S in Example 4.2 to obtain the low-frequency and the high-frequency components.

Relevant MATLAB code that generates weighting function S, and filtered profiles w and r is shown below. Profile w is plotted along with profile z in Fig. 4.2a. Profile r is shown in Fig. 4.2b.

```
n=size(z,1);
x=(0:1:n-1)'*dx;
w=conv(z,S);              % convolve profile and filter
q=alpha/dx;               % q is half the length of the filter
w=w(q+1:size(w,1)-q); % extract relevant portions only
plot(x,z,'r',x,w,'b'); % plot z and w
xlabel('Distance (mm)');
ylabel('Amplitude (\mum)');
r=z - w;                  % Roughness profile
plot(x,r,'k');            % plot profile r
xlabel('Distance (mm)');
ylabel('Amplitude (\mum)');
```

It is evident from Fig. 4.2a that the filtered profile w captures the general shape of the longer wavelength component of the input profile z. There is however some distortion in w (it resembles a triangular wave with ups and downs and is not a nice smooth sinusoidal curve), which therefore also affects profile r. We explain the origin of this distortion in Example 4.4.

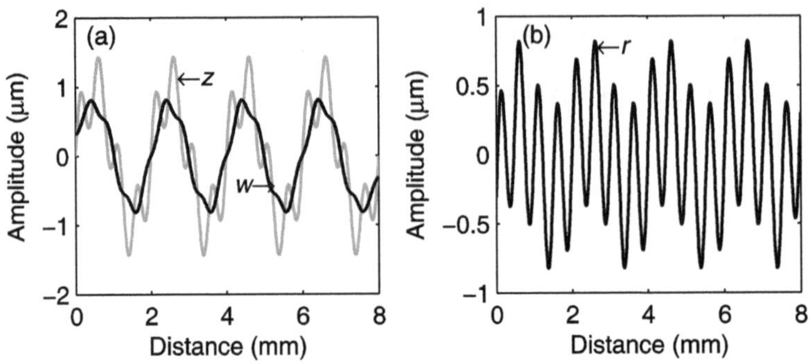

Fig. 4.2 **a** Profiles z and w shown. Profile w generated using a rectangular weighting function of size 0.8 mm. **b** Profile r

4.2 Relationship Between Frequency-Domain and Time-Domain Filtering

In the frequency-domain technique discussed previously, we knew exactly what wavelengths were transmitted and what wavelengths were suppressed. That is because we defined the filter in the frequency domain itself. In the averaging method however, we did not clearly specify the relationship between width of a window and wavelength cutoff. Does the rectangular filter S in Example 4.2 of width 0.8 mm necessarily have a cutoff of 0.8 mm? The answer is no. The width of a window in time/spatial domain is not in itself the value of a wavelength cutoff. Further, a rectangular filter in the spatial domain is not the same as a step filter in the frequency domain.

A filter can be defined in either the spatial domain or the frequency domain, and the two definitions are related to each other by the Fourier transform and the inverse Fourier transform. Therefore, the rectangular filter in the spatial domain will have a rather different shape in the frequency domain as determined by its Fourier transform, and the step filter in the frequency domain will have a different shape in the spatial domain as determined by its inverse transform. These concepts are illustrated in the examples below.

Example 4.4. Plot the transmission characteristics of the rectangular filter S.

The transmission characteristics of the rectangular weighting function can be obtained by taking the DFT of the weighting function. However, recall that the length of the weighting function S is only 801. Performing the DFT of this weighting function will provide information on wavelengths ranging from the Nyquist frequency (twice the spacing; equal to 2 µm in this example) to a maximum wavelength equal to the length of the signal (0.8 mm). In order to capture the response to a broader range of wavelengths up to the length of the profile itself (8 mm), the weighting function length has to be increased to the length of the profile. This can be done by zero-padding the weighting function array so that its length is also 8,000. Zero-padding may be done in many ways; we can zero pad at the beginning, at the end, or

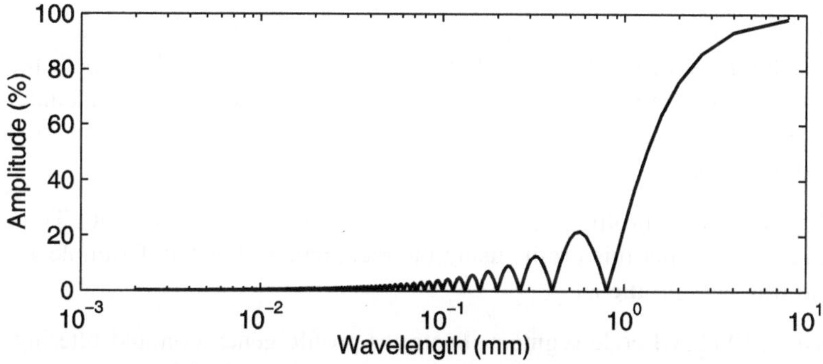

Fig. 4.3 Transmission characteristics of rectangular weighting function *S*

symmetrically on both sides. Zero-padding in different ways will result in different phase offsets at the output. For now, we will zero pad equally on both sides and study the effect of this on the phase in Example 4.6.

The following MATLAB segment illustrates the process of zero-padding and transformation to the frequency domain. Figure 4.3 shows the DFT of the weighting function. It is customary to plot the *x* axis of the transmission characteristics on a logarithmic scale. Notice that a rectangular weighting function of 0.8 mm width does not suppress all wavelengths below 0.8 mm completely. The attenuation of the amplitude is a function of the wavelength of the input sinusoid. Profile z_1 (amplitude 1 μm and wavelength 2 mm) will emerge from the filter with the same wavelength, but with 75.63% of the input amplitude or 0.7563 μm. Profile z_2 (amplitude of 0.5 μm and wavelength of 0.5 mm) will emerge from the filter with an amplitude of only 18.86% of the input or 0.0943 μm.

```
S=[zeros(3600,1);S; zeros(3599,1)];% center the rectangular array
Sf=fft(S);                         % take the FFT
j=(2:1:floor(n/2)+1)';             % create the wavelength array
wave=n*dx./(j-1);
semilogx(wave,100*abs(Sf(2:floor(n/2)+1,1))); % plot amplitude
xlabel('Wavelength (mm)');
ylabel('Amplitude (%)');
```

The transmission plot shown in Fig. 4.3 is not smooth. Because of the many humps, the filter transmits some frequencies while blocking others that are fairly close. This explains the origin of the distortion in Example 4.3: the filter did not completely suppress the smaller wavelength component, but allowed a small fraction to leak through.

4.3 Filtering Profiles: Putting It All Together

We have discussed a variety of different topics in this and in Chap. 3. Let us attempt to put it all together with the following examples. We will generate a profile *z* that contains an odd number of sinusoids, and then attempt to filter this profile using a rectangular filter defined in the time domain. Then, we will transform both the filter and profile, and perform filtering in the frequency domain. We will encounter a phase problem in the frequency domain that is tied to the wrap-around effect and our zero-padding strategy.

Example 4.5. Generate a profile z (8 mm long, 1 μm spacing) of wavelength 3 mm and amplitude 1 μm. Filter this profile using the rectangular filter S in Example 4.2 to obtain the waviness profile w_1.

The following MATLAB code segment illustrates profile generation and filtering. The results are plotted in Fig. 4.4a. We continue from Example 4.2 where filter S was defined.

```
n=8000;
dx=0.001;
x=(0:dx:n*dx-dx)';
z=sin(2*pi.*x/3); % generate z
plot(x,z)
q=(size(S,1)-1)/2;
w1=conv(S,z);       % convolve z and S
w1=w1(q+1:n+q);     % extract the central n points
plot(x,z,'b-',x,w1,'r-')
```

Example 4.6. Use the transform of filter S (Example 4.4) to filter profile z (Example 4.5) in the frequency domain to obtain waviness profile w_2.

The following MATLAB code segment illustrates filtering in the frequency domain. The results are plotted in Fig. 4.4a, b.

```
S=[zeros(3600,1); S; zeros(3599,1)];% center filter in S
Sf=fft(S);                          % take the FFT of the filter
zf=fft(z);                          % take the FFT of z
w2=ifft(zf.*Sf);                    % term by term multiplication
                                    % and inverse FFT

plot(x,z,'b-',x,w1,'r-',x,w2 ,'g-')
w2=[w2(4001:8000,1); w2(1:4000)];
plot(x,z,'b-',x,w1,'r-',x,w2,'g-')
```

Notice that the filtered profile w_2 is distinctly out of phase (see Fig. 4.4a) with the input, unlike profile w_1. In fact, it is out of phase by exactly half the profile length. Therefore, if the first half of profile w_2 were extracted and pasted to the end of profile w_2, and plotted as in Fig. 4.4b, the result would be identical to that of w_1.

This phase offset is the result of our zero-padding strategy of equally distributing zeros on both sides of the filter combined with the wrap-around effect. Recall that in the frequency domain, the profile is considered to be periodic, that is, wrapped around itself (as if it were a roundness profile). Also remember that the filter window is of same length as the profile. At the start of convolution, the window is positioned at the left extreme of what is considered as the first profile point. However, because the profile is wrapped around, the window overlaps the second half of the profile. Stepping the window to the right results in a moving average performed on the

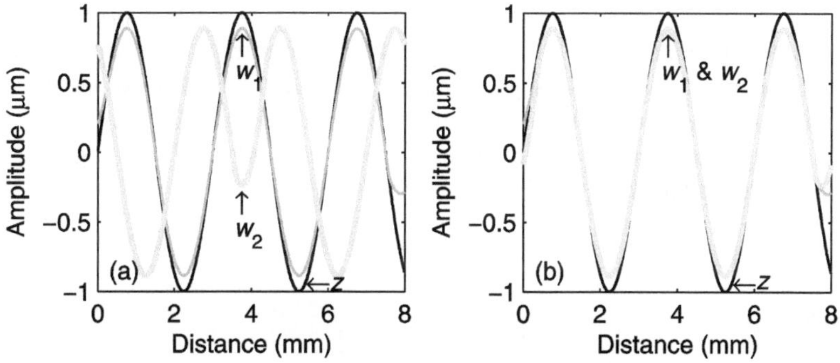

Fig. 4.4 **a** Profiles z, w_1 (filtered profile, time-domain convolution) and w_2 (filtered profile, frequency-domain method, before phase correction). **b** Profiles z, w_1 and w_2 (after phase correction)

second half of the profile. Gradually, as the window steps further to the right, it sees the first half of the profile. Determination of the DFT would be analogous to computing the first n filtered points in this manner and will result in a filtered profile with the observed phase offset.

MATLAB's `conv` assumes the profile is open and therefore we do not encounter this phase problem. We can solve this wrap-around and phase problem by zero-padding both the profile and the filter as shown. Notice that w_3 is now identical to w_1.

```
z2(16000,1)=0;        % generate a~longer array z2
z2(4001:12000,1)=z;  % assign the first 12000 points
S2(16000,1)=0;        % generate a~longer array S2
S2(4001:12000,1)=S;  % assign the first 12000 points
Sf2=fft(S2);
zf2=fft(z2);
w3=ifft(zf2.*Sf2);    % term by term multiplication
                      % and inverse FFT
w3=[w3(12001:16000,1);w3(1:4000,1)];
plot(x,z,'b-',x,w1,'r-',x,w3 ,'g-')
```

4.4 Summary

In this chapter we have discussed an important result from signal processing without actually mentioning it. If we review our discussion, we notice the following:

1. In time-domain filtering, we define a filter (S) in the time domain (such as a rectangular filter), convolve it (moving average) with the profile (z), which also resides in the time domain, and directly obtain the filtered output (w).
2. In frequency-domain filtering, we define a filter in the frequency domain (Sf), evaluate the transform of the raw profile (zf), and multiply zf and Sf to obtain the transform of the filtered profile (wf). Inverse transform yields the desired output in the time domain (w).

The filter S in the time domain and its counterpart Sf in the frequency domain are Fourier transform pairs; z and zf, w and wf are also Fourier transform pairs. From the above, it appears that multiplication in the frequency domain produces identical results with convolution in the time domain. This is indeed true, and is discussed in detail in many signal processing textbooks (Oppenheim *et al.*, 1999; Ingle and Proakis 1996; McClellan *et al.*, 1997).

Exercises

4.1. In convolution of a profile containing n points with a weighting function containing m points, what is the length of the filtered output that contains distortion-free data? Distortion-free data can be obtained when the entire weighting function is inside the profile. Near the edges on either side, some elements of the weighting function do not see corresponding profile elements and therefore, the weighted average is not correctly estimated. Propose solutions to compensate for this problem.

4.2. Write a function that performs a moving average of two arrays A and B without using MATLAB's `conv` function. Now, use this function to perform convolution of the rectangular weighting function S defined in Example 4.2 and the profile z defined in Example 3.1. Compare the speed of this function against MATLAB's `conv` function (use functions `tic` and `toc`). Which function is faster and why?

4.3. Generate a sinusoidal profile of amplitude 1 μm and wavelength 0.1 mm defined in the interval $0 \leqslant x < 8$ mm with a sampling interval $\Delta x = 1$ μm. Convolve this profile with the rectangular weighting function S defined in Example 4.2. What is the amplitude and wavelength of the output sinusoidal profile? Repeat the process for different wavelengths from 0.1 mm to 4 mm in intervals of 0.01 mm (write a program that will perform this). Plot the amplitude of the output sinusoid as a function of the wavelength of the input sinusoid. (Hint: the plot must resemble that shown in Fig. 4.3. Decrease the interval further if necessary.)

4.4. Generate and plot the amplitude and phase transmission characteristics of a triangular filter S_t, defined as shown in the spatial domain in the interval $-\lambda c \leqslant x \leqslant \lambda c$, where $\lambda c = 0.8$ mm and $\alpha = 0.4429$

$$
\begin{aligned}
S_t &= 0\,, && -\lambda c \leqslant x < -\alpha\lambda c \\
&= (1/\alpha\lambda c)^2 x + (1/\alpha\lambda c)\,, && -\alpha\lambda c \leqslant x < 0 \\
&= -(1/\alpha\lambda c)^2 x + (1/\alpha\lambda c)\,, && 0 \leqslant x < \alpha\lambda c \\
&= 0\,, && \alpha\lambda c \leqslant x \leqslant \lambda c
\end{aligned}
$$

4.5. Generate a sinusoidal profile z of amplitude 1 μm and wavelength 0.7 mm, at a sampling interval of 1 μm, over a length of 8 mm. Convolve this profile with filter S_t (Exercise 4.4) to obtain the filtered profile. What is the amplitude of the filtered profile?

4.6. Zero-pad the filter S_t generated in Exercise 4.4, so that the triangle is symmetrically placed in the array. Then, perform the DFT of the filter and estimate the amplitude transmission of profile z from Exercise 4.5 simply by observing the transmission characteristics of the filter. Does it agree with that estimated in Exercise 4.5?

4.7. Multiply the DFT of profile z from Exercise 4.5 with the DFT of the zero-padded filter S_t from Exercise 4.4. Perform the inverse DFT to obtain the filtered profile. Compensate for the phase (as in Example 4.6). Compare this profile against that obtained in Exercise 4.5. Does it agree identically?

4.8. Compensate for the wrap-around effect in Exercise 4.7 by zero-padding (as in Example 4.6). Now, does the filtered profile identically agree with the result from Example 4.5?

References

Ingle, V.K. and Proakis, J.G. 1996, *Digital Signal Processing Using MATLAB V.4*, PWS Publishing Company. Boston, MA, USA.

McClellan, J.H., Burrus, C.S., Oppenheim, A.V., Parks, T.W., Schafer, R.W. and Schuessler, H.W. 1997, *Computer-Based Exercises for Signal Processing Using MATLAB 5*, Prentice Hall. Upper Saddle River, NJ, USA.

Oppenheim, A.V., Schafer, R.W. and Buck, J.R. 1999, *Discrete-Time Signal Processing*, 2nd edn, Prentice Hall. Upper Saddle River, NJ, USA.

Chapter 5
Gaussian Filter

5.1 Introduction

The most widely used filter for surface profile analysis is the Gaussian filter. It is described both in the American Standard (American Society of Mechanical Engineers 2002) and the international Standard (International Organization for Standardization 1996). The weighting function (the definition of the filter in time/spatial domain) of the Gaussian filter is given by

$$S(x) = \frac{1}{\alpha\lambda c}\exp\left[-\pi\left(\frac{x}{\alpha\lambda c}\right)^2\right], \tag{5.1}$$

where $\alpha = \sqrt{\ln 2/\pi} = 0.4697$, x is the position from the origin of the weighting function and λc is the long-wavelength roughness cutoff.

The transmission characteristics can be obtained by performing the Fourier transform of the continuous function $S(x)$, as shown below

$$Sf(\lambda) = \int_{-\infty}^{\infty} S(x)\mathrm{e}^{i\lambda x}dx = \int_{-\infty}^{\infty} \frac{1}{\alpha\lambda c}\mathrm{e}^{\left[-\pi\left(\frac{x}{\alpha\lambda c}\right)^2\right]}\mathrm{e}^{i\lambda x}dx = \mathrm{e}^{\left[-\pi\left(\alpha\frac{\lambda c}{\lambda}\right)^2\right]}. \tag{5.2}$$

From Eq. 5.2, it can be seen that the amplitude attenuation of a sinusoid whose wavelength is equal to the cutoff ($\lambda = \lambda c$) is 0.5. Thus, the filter transmits 50% at the cutoff. The following examples illustrate the determination of the filter weighting function and transmission curve of the Gaussian filter.

Example 5.1. Plot the Gaussian filter S in spatial domain in the interval $-\lambda c \leqslant x \leqslant \lambda c$. Let $\lambda c = 0.8$ mm and sampling interval $\Delta x = 1$ µm.

The following MATLAB code generates and plots (see Fig. 5.1a) the filter in the desired interval:

```
lambdac=0.8; % in mm
dx=0.001;    % in mm
```

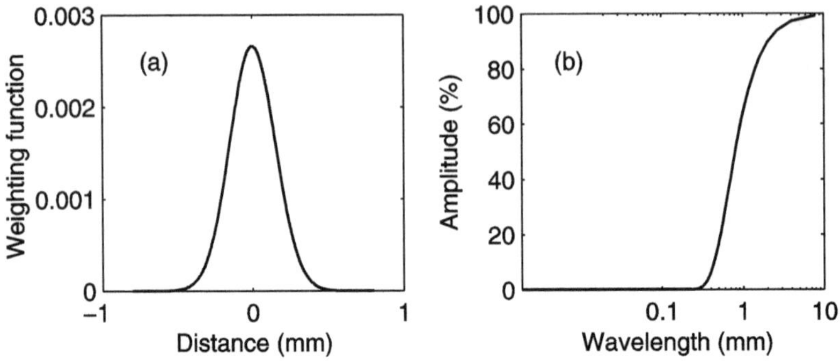

Fig. 5.1 **a** Plot of the Gaussian filter; $\lambda c = 0.8$ mm. **b** Transmission characteristics of the Gaussian filter

```
x=(-lambdac:dx:lambdac)';
alpha=0.4697;
S=(1/(alpha*lambdac)).*exp(-pi*(x/(alpha*lambdac)).^2);
            % generate the Gaussian filter
S=S/sum(S); % normalize to unit sum
plot(x,S);
xlabel('Distance (mm)');
ylabel('Weighting function');
```

Example 5.2. Generate the amplitude transmission characteristics of the Gaussian filter S and evaluate the amplitude transmission of a sinusoid whose wavelength is equal to the cutoff ($\lambda c = 0.8$ mm).

The transmission characteristic of the Gaussian filter can be obtained as described for the rectangular filter in Example 4.4 in Chap. 4. It is plotted in Fig. 5.1b. The amplitude attenuation of a sinusoidal component of wavelength 0.8 mm is 0.5 (11th element of array Sf). Thus, the Gaussian filter transmits 50% at the cutoff. Notice that the transmission plot in Fig. 5.1b is not sharp at the cutoff, but that there is a finite slope. This implies that the filter does not abruptly suppress all wavelengths smaller than the cutoff, but gradually attenuates wavelengths near the cutoff. The transmission characteristics can also be plotted by directly sampling Eq. 5.2.

```
m=size(S,1); % length of Gaussian filter
l=8;         % length of a profile is assumed to be 8 mm
n=l/dx;      % number of profile points
S=[zeros(n/2-floor(m/2),1); S; zeros(n/2-floor(m/2)-1,1)];
% center the filter and zero-pad to 8000 long
Sf=fft(S);   % DFT of S
j=(2:1:floor(n/2)+1)';% generate wavelength array for X axis of
                      % transmission plot
wave=n*dx./(j-1);
semilogx(wave,100*abs(Sf(2:floor(n/2)+1,1)));
xlabel('Wavelength (mm)');
ylabel('Amplitude (%)');
```

5.2 High-Pass and Low-Pass Filters

The filter described thus far is referred to as a low-pass filter because it transmits low frequencies only (wavelengths greater than the cutoff). The rectangular and step filters described in Chaps. 3 and 4 are also low-pass filters. A low-pass filter is an averaging filter in that it will produce a smoother profile. A filter can also be designed to transmit only high frequency signals. Such a filter is known as a high-pass filter. The transmission characteristics of the high-pass Gaussian filter is given by

$$Sf(\lambda) = 1 - \mathrm{e}^{\left(-\pi\left(\alpha\frac{\lambda c}{\lambda}\right)^2\right)}. \tag{5.3}$$

The high-pass Gaussian filter also transmits 50% at the cutoff. This is an important consequence of the complementary definitions of the Gaussian low-pass and high-pass transmission characteristics. It can be shown that the high-frequency signal in the spatial domain is simply the difference of the low-frequency signal from the raw profile. The application of two distinct filters, a high-pass and a low-pass, to capture high- and low-frequency signals is therefore not necessary.

5.3 Roughness, Waviness, and Form Using the Gaussian Filter

Whether the filtered output is termed roughness, waviness or form depends on the type of filter (high-pass or low-pass) and the values of the cutoffs. As mentioned in Chap. 3, because roughness, waviness, and form represent a certain defined wavelength bandwidth, two cutoffs are required to properly establish each band, one for the lower bound and another for the upper bound. The ASME B46.1 Standard specifies cutoff λcw as the waviness long-wavelength cutoff, λsw as the waviness short-wavelength cutoff, λc as the roughness long-wavelength cutoff and λs as the roughness short-wavelength cutoff. In general, λc is chosen to be equal to λsw. Both the ASME B46.1 and the ISO 11562 Standards have preferred values for the cutoffs. Further, they provide guidelines on cutoff ratios that may be chosen as well. For example, selecting λc and an appropriate ratio for $\lambda c/\lambda s$ will then limit the size of the stylus tip that can be chosen (American Society of Mechanical Engineers 2002).

The filter associated with the cutoff λs is a low-pass filter while that associated with λc is a high-pass filter. The pair of filters applied on the profile produces roughness. Waviness can be obtained by applying the pair of filters corresponding to λsw (low-pass) and λcw (high-pass). Form is generally regarded as comprising of all large-wavelength components and therefore a cutoff λf at the lower end is generally regarded as sufficient. Note that the ASME B46.1 Standard does not define λf. Figure 5.2 illustrates the different filters described above.

From the above, it appears that the determination of roughness (or waviness) requires two filtering operations. In practice, however, a few simplifications permit the determination of roughness (or waviness) from a single filtering operation. First, the

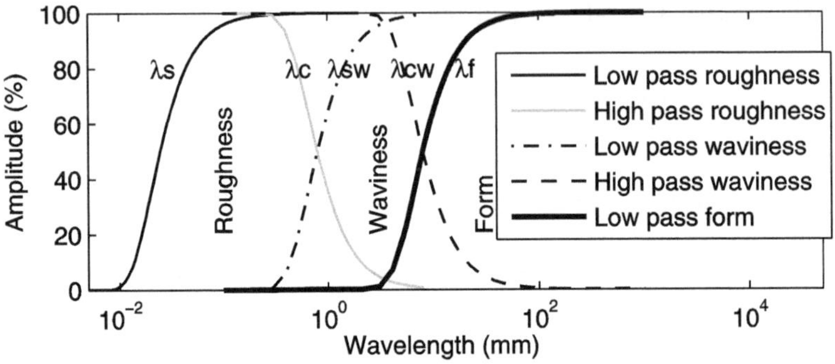

Fig. 5.2 Transmission characteristics of five filters that partition the profile into roughness, waviness, and form ($\lambda s = 25$ µm, $\lambda c = 0.8$ mm, $\lambda sw = 0.8$ mm, $\lambda cw = 8$ mm, and $\lambda f = 8$ mm)

low-pass filter with λs cutoff is sometimes ignored or the instrument automatically filters the raw data. Second, the profile length is often not long enough to capture form and therefore the high-pass filter with cutoff λcw is not applied. And finally, as mentioned in Sect. 5.2, the complementary definitions of the Gaussian high- and low-pass filters permit the evaluation of roughness and waviness using one filter, either the high-pass or the low-pass with cutoff λc.

In summary, the roughness profile can be obtained by filtering the raw profile using a Gaussian high-pass filter with cutoff λc. The waviness profile is simply the difference between the raw profile and the roughness profile. The application of the Gaussian low-pass filter (λsw) on the raw profile will result in the waviness profile. In that case, the roughness profile is the difference between the raw profile and the waviness profile. Needless to say, the two methods yield identical results.

5.4 Effect of Cutoff

As λc increases, a larger bandwidth is allowed into the roughness profile, thus increasing amplitude parameters such as *Ra*, computed on the roughness profile. On the other hand, the same cutoff will result in some wavelength content being removed from the waviness profile resulting in a smaller amplitude parameter *Wa* computed on the waviness profile.

5.5 Phase Characteristics

While we have studied the amplitude transmission characteristics of the Gaussian filter in considerable detail, the phase characteristics have not been mentioned yet. The Gaussian filter does not introduce any relative phase shift between sinusoids of different wavelengths. In other words, every sinusoid entering the filter will emerge from the filter in phase with the input profile. This zero-phase property is one of the key advantages of the Gaussian filter.

5.6 Summary

The Gaussian filter is probably the most widely used filter today. We reviewed its definition and implementation in this chapter. We also discussed the definition of a high-pass and a low-pass filter. Also we discussed how Gaussian high-pass and low-pass filters are complementary, thus rendering two filtering operations unnecessary to obtain roughness and waviness. The Gaussian filter has zero phase and therefore there are no distortions in the output.

Exercises

5.1. Generate the transmission characteristic curve for the Gaussian low-pass filter by sampling the continuous function in Eq. 5.2. Assume $\lambda c = 0.8$ mm and sampling interval $\Delta x = 1$ µm. Is the curve identical to that obtained in Fig. 5.1b?

5.2. Generate the plot shown in Fig. 5.2, assuming the profile is 1,000 mm long, using a spacing parameter $\Delta x = 1$ µm, with the cutoff values as shown in Fig. 5.2.

5.3. Write a function `myGaussian.m`, that accepts two arguments – spacing and cutoff (both in identical units) – and generates the corresponding low-pass spatial domain Gaussian filter that is normalized so that the sum of all weights is unity.

5.4. From the amplitude transmission curve for the Gaussian filter, what is the amplitude attenuation of a sinusoid z_1 of 2 mm wavelength and 1 µm amplitude when it is filtered using a low-pass Gaussian filter of cutoff 0.8 mm. Use spacing parameter $\Delta x = 1$ µm. What is the amplitude attenuation of a sinusoid z_2 of 0.5 mm wavelength and 0.5 µm amplitude when it emerges from the same filter?

5.5. From Exercise 5.4 and from knowledge of the phase characteristics of the Gaussian filter, compute and plot the filtered output generated when this Gaussian filter is applied to a profile that consists of the sum of the two sinusoidal components discussed in Exercise 5.4.

5.6. Verify the result of Exercise 5.4 using spatial-domain convolution. Use the Gaussian filter function from Exercise 5.3 to generate the filter.

5.7. Verify the result of Exercise 5.4 using frequency-domain multiplication. Zero-pad the Gaussian filter generated in spatial domain so that the window is symmetrically placed in the array of length 8,000. Then, perform a term-by-term multiplication of the DFTs of the filter and the profile, evaluate the inverse DFT and extract the relevant portions to compensate for phase (see Example 4.6). This waviness profile must be identical to that produced in Exercises 5.5 and 5.6, except maybe at the edges.

5.8. Use the spatial-domain convolution method to filter profile z in Example 3.1, using the low-pass Gaussian filter at three different cutoffs: 0.5 mm, 0.8 mm, and 2 mm. Discard one half of the cutoff length from the beginning and end of w. Compute Wa on the filtered waviness profile w, where Wa is defined as $Wa = \frac{\sum_{i=1}^{n} |w(i)|}{n}$, and $w(i)$ is the ith element of the waviness array and n is the length of array w. Does Wa increase or decrease with increasing cutoff and why?

References

American Society of Mechanical Engineers 2002, *ASME B46.1-2002, Surface Texture (Surface Roughness, Waviness and Lay)*, ASME. New York, NY, USA.

International Organization for Standardization 1996, *ISO 11562:1996, Geometrical Product Specification (GPS) – Surface Texture: Profile Method – Metrological Characteristics of Phase Correct Filters*, ISO. Geneva, Switzerland.

Chapter 6
The 2RC Filter

6.1 Introduction

The earliest filter used in surface metrology was the 2RC filter. It was first implemented in hardware, but is now conveniently implemented in software. The 2RC filter does not have a linear phase and there is a need for separate roughness and waviness filters. The 2RC, although used from the early days of surface profile filtering, is discussed after the Gaussian filter because the filter poses more challenging implementation issues. The ASME B46.1 Standard (American Society of Mechanical Engineers 2002) discusses the 2RC filter but the international Standards do not.

6.2 The 2RC High-Pass Filter

Historically, the 2RC filter (Whitehouse and Reason 1965) was used to obtain only roughness from a raw profile, and therefore the reported literature typically only provides a weighting function for the 2RC roughness filter. In order to obtain the waviness profile, a separate weighting function is required (see Exercise 6.1). The weighting function for the high-pass roughness filter is given by

$$S(x) = \frac{A}{\lambda c}\left(2 - A\frac{|x|}{\lambda c}\right)\exp\left(-A\frac{|x|}{\lambda c}\right), \qquad (6.1)$$

where $A = 3.64$, x is the position from the origin of the weighting function and λc is the long-wavelength roughness cutoff. The transmission characteristics of the high-pass filter can be obtained from the Fourier transform of the continuous function $S(x)$:

$$Sf(\lambda) = \left(1 - jk\frac{\lambda}{\lambda c}\right)^{-2}, \qquad (6.2)$$

where $k = 1/\sqrt{3}$ and $j = \sqrt{-1}$. There is an important point to note regarding the weighting function in Eq. 6.1. Although the purpose of the weighting function is to provide the high-frequency component of the profile, spatial-domain convolution is essentially an averaging operation. The result of convolving the weighting function defined in Eq. 6.1 will only provide a "mean line". The difference between this mean line and the profile yields the roughness profile. This is not evident from Eq. 6.1. However, for the more interested reader, the derivation in Sect. 6.3 provides a clue as to why this is so.

The amplitude attenuation of the 2RC filter is 75% for a sinusoid whose wavelength is equal to the cutoff. This follows from Eq. 6.2 when $\lambda = \lambda c$. Also, the filter does not have constant phase characteristics. This implies that sinusoids of different wavelengths will be offset by different amounts in phase from the input. This introduces a distortion in the profile and is not a desirable feature. The 2RC filter was used for many years despite this limitation, because digital filters were not realized at the time of the origin of surface profilometry. The following examples illustrate these aspects of the 2RC filter.

Example 6.1. Plot the weighting function, amplitude transmission, and phase characteristics of the high-pass 2RC filter. Assume cutoff $\lambda c = 0.8$ mm.

The weighting function of the 2RC filter is generated in the interval $-2\lambda c \leqslant x \leqslant 0$ because the filter is not symmetrical about the origin. In fact, the filter is one-sided with the leading edge at the origin.

The transmission characteristic can be generated in multiple ways. In this example, we will directly sample Eq. 6.2. Figure 6.1a–c shows the weighting function, the amplitude transmission, and phase characteristics of the 2RC filter.

```
lambdac = 0.8;                  % cutoff in mm
dx = 0.001;                     % spacing in mm
A = 2*pi/sqrt(3);
x = (-2*lambdac:dx:0)';
for j=1:1:length(x)             % generate weighting function
    S(j,1)=(A/lambdac)*(2-(A/lambdac)*abs(x(j)))*
    exp(-(A/lambdac)*abs(x(j)));
end
S = S/sum(S);                   % normalize to unit sum
plot(x,S)
xlabel('Distance (mm)');
ylabel('Weighting function');
lamda = (0.002:0.001:8)';       % generate wavelength axis
Sf = (1-i*1/sqrt(3).*(lamda/lambdac)).^(-2);
                                % sample the continuous
                                % function in Eq. 6.2
semilogx(lamda,100*abs(Sf)); % take absolute values to determine
                                % amplitude transmission
xlabel('Wavelength (mm)');
ylabel('Amplitude (%)');        % plot amplitude as %
semilogx(lamda, angle(Sf));     % determine angle to plot phase
xlabel('Wavelength (mm)');
ylabel('Phase (radians)');      % plot phase in radians
```

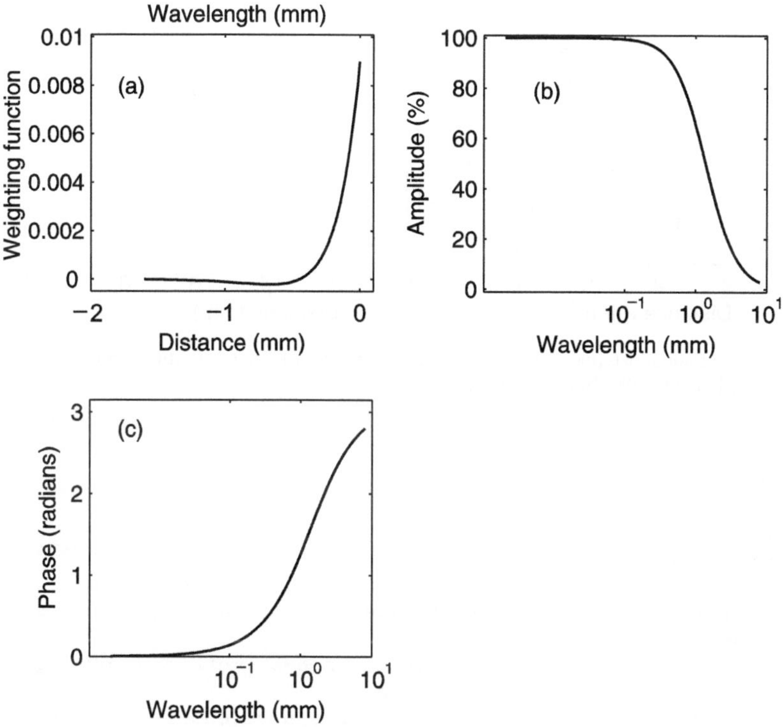

Fig. 6.1 **a** Weighting function of two-resistor-capacitor (2RC) filter. **b** Amplitude transmission characteristics. **c** Phase characteristics

Example 6.2. Filter the profile z generated in Example 3.1 with a 2RC high-pass filter of 0.8 mm cutoff using spatial-domain convolution.

The 2RC high-pass filter is not symmetrical about the origin. Therefore the filter has to be mirrored about the origin prior to convolution with the profile (recall that the definition of convolution involves a mirroring operation; this mirroring offsets the in-built mirror in `conv`). Also, the result of convolution is a profile of length $n+m-1$, where n is the length of the raw profile and m is the length of the filter array. In the case of symmetrical filters, the central n elements from the convolution array represent the waviness profile, but in the case of the 2RC filter, the first n elements represent the mean line. This is because the origin in the filter array (the head of the filter) carries the greatest weight and commences filtering as soon as it sees the first profile point. Towards the end of the profile, as the filter recedes from the surface, the head of the filter is outside the profile and no longer evaluates a true moving average. In the case of the Gaussian filter, the point with greatest weight is symmetrically placed in the middle and, therefore, the central points of the convolution array are extracted as the waviness profile. Figure 6.2a shows the raw profile and the mean line. Figure 6.2b shows the roughness profile.

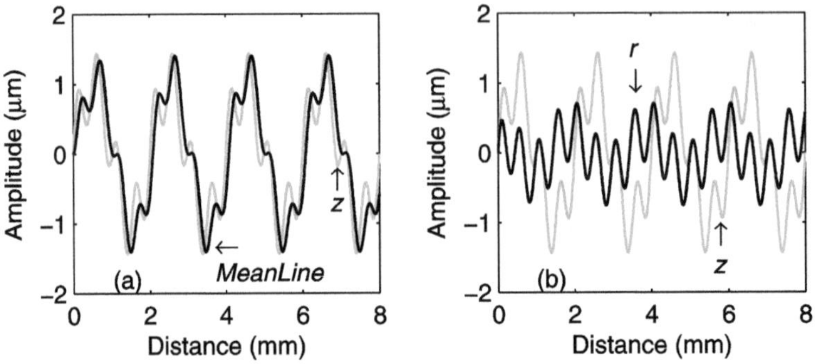

Fig. 6.2 **a** Profile z and mean line generated using a 2RC high-pass filter. **b** Profile z and roughness profile generated using a 2RC high-pass filter

```
dx = 0.001;
l = 8;
n = l/dx;
x = (0:dx:l-dx)';
z1 = 1*sin(2*pi.*x/2);        % amplitude 1 um, wavelength 2 mm
z2 = 0.5*sin(2*pi.*x/0.5); % amplitude 0.5 um, wavelength 0.5 mm
z = z1 + z2;
S = flipud(S);                % mirror the non-symmetric weighting
                              % function
C = conv(S, z);               % conv performs a second flip before
                              % moving average
MeanLine = C(1:n);            % extract the first n points
r = z-MeanLine;
plot(x,z,'r',x,MeanLine,'b');
xlabel('Distance (mm)');
ylabel('Amplitude (\mum)');
plot(x,z,'r',x,r ,'b');
xlabel('Distance (mm)');
ylabel('Amplitude (\mum)');
```

6.3 More on the 2RC Filter

Early implementations of the 2RC filter were in real time. As the instrument traversed the surface, profile points were fed to a 2RC network as voltage signal (Vin). The filtered output was collected as a voltage signal ($Vout$) and later converted to units of height. The transfer function ($Hf_1 = Vout/Vin$) for a one-stage RC filter, from any standard electrical engineering textbook, is given by

$$Hf_1 = \frac{jRC\omega}{1 + jRC\omega}, \tag{6.3}$$

where ω is the frequency given by $\omega = \frac{2\pi}{\lambda}$, $j = \sqrt{-1}$, and R and C are the resistors and capacitors. The transfer function is the transmission characteristics of the filter and is defined in the frequency domain (hence, according to our adopted convention, the f in the variable Hf). For the two-stage network as shown in Fig. 6.3, the transfer function is simply the square of the transfer function of the one-stage network. Thus,

$$Hf_2 = \frac{V_{\text{out}}}{V_{\text{in}}} = Hf_1^2 = \left(\frac{jRC\omega}{1+jRC\omega}\right)^2 . \tag{6.4}$$

The definition of the 2RC filter specifies a transmission of 75% at the cutoff λc. The value of RC that satisfies this condition can be obtained as follows. Replacing $\omega = \frac{2\pi}{\lambda c}$ in Eq. 6.4, we get

$$|Hf_2| = \frac{(2\pi RC/\lambda c)^2}{\sqrt{\left(1-(2\pi RC/\lambda c)^2\right)^2 + (2\pi RC/\lambda c)^2}} = 3/4 . \tag{6.5}$$

Expanding and solving,

$$RC = \frac{\lambda c\sqrt{3}}{2\pi} . \tag{6.6}$$

In order to implement the filter in spatial domain, we require an expression for the weighting function. This can be obtained from the inverse transform of the transfer function. Let us rewrite the transfer function as

$$Hf_2 = \left(1 - \frac{1}{j\omega RC}\right)^2 = 1 - \frac{2}{RC}\frac{1}{\frac{1}{RC}+j\omega} + \frac{1}{(RC)^2}\frac{1}{\left(\frac{1}{RC}+j\omega\right)^2} . \tag{6.7}$$

Taking inverse Fourier transform, we get

$$H(t) = \delta(t) - \frac{1}{RC}\left(2 - \frac{t}{RC}\right)e^{\frac{-t}{RC}} , \tag{6.8}$$

where $\delta(t)$ is an unit impulse.

$H(t)$ is the impulse response of the 2RC filter. It has two parts, the unit impulse and the time-varying exponential part. The time-varying part generates the DC mean line and the impulse part is the instantaneous AC part. The weighting function for the DC component can be obtained by substituting the value of RC in the equation above. Thus,

$$S(x) = \frac{A}{\lambda c}\left(2 - \frac{xA}{\lambda c}\right)e^{\frac{-xA}{\lambda c}} , \quad 0 \leqslant x < \infty , \tag{6.9}$$

where $A = \frac{2\pi}{\sqrt{3}}$.

Fig. 6.3 A 2RC electrical network. C Capacitors, R resistors, V_{in} input voltage signal, V_{out} output voltage signal

The impulse response answers a question posed early on as to why the weighting function of the high-pass 2RC filter only provides a mean line. As seen here, the weighting function only captures the DC component of the impulse response. The instantaneous AC component (roughness profile) is the difference between the raw profile and the DC mean line.

As a final note on the 2RC filter, the transfer function with the 75% condition is obtained by substituting for *RC* in

$$Hf_2 = \left(\frac{jRC\omega}{1+jRC\omega}\right)^2 = \frac{1}{\left(1-\frac{j\lambda}{\lambda c\sqrt{3}}\right)^2}. \tag{6.10}$$

6.4 Comparison of the 2RC and Gaussian Filters

An example surface profile (d_med – Type D roughness specimen) can be downloaded from the National Institute of Standards and Technology's (NIST) Internet Based Surface Metrology Algorithm Testing System Website (The National Institute of Standards and Technology; Bui and Vorburger 2007). Header and footer information (the first 15 and last 2 lines) is deleted prior to loading into MATLAB. This profile is filtered using a Gaussian low-pass filter at 0.8 mm cutoff. The same profile is also filtered using a 2RC high-pass filter at the same cutoff. The raw profile and the Gaussian waviness profiles are plotted in Fig. 6.4a. Notice that the waviness profile follows the peaks and valleys with no phase distortion. The corresponding roughness profile is shown in Fig. 6.4 b.

The raw profile and the mean line (not waviness profile) generated using a 2RC filter at 0.8 mm cutoff are shown in Fig. 6.4c. Notice that the mean line is shifted in phase. The roughness profile in Fig. 6.4d therefore shows a distortion (artificial features) because of this phase offset.

Also notice that the roughness profile in Fig. 6.4d is generally of larger amplitude than that in Fig. 6.4b, although both were generated using the same cutoff. This is because the 2RC filter transmits 75% at the cutoff, while the Gaussian filter only transmits 50% at the cutoff. Therefore, wavelengths at or near the cutoff will be transmitted by the 2RC filter with limited attenuation, hence the larger amplitudes.

6.5 Summary

The 2RC filter was the earliest filter used in surface metrology. It was originally implemented using real-time electrical networks but is now conveniently implemented digitally. The filter suffers from several drawbacks: (a) it introduces a phase distortion in the roughness profile, (b) there is a need for separate roughness and waviness filters, and (c) there is edge distortion. The Gaussian filter was introduced later to

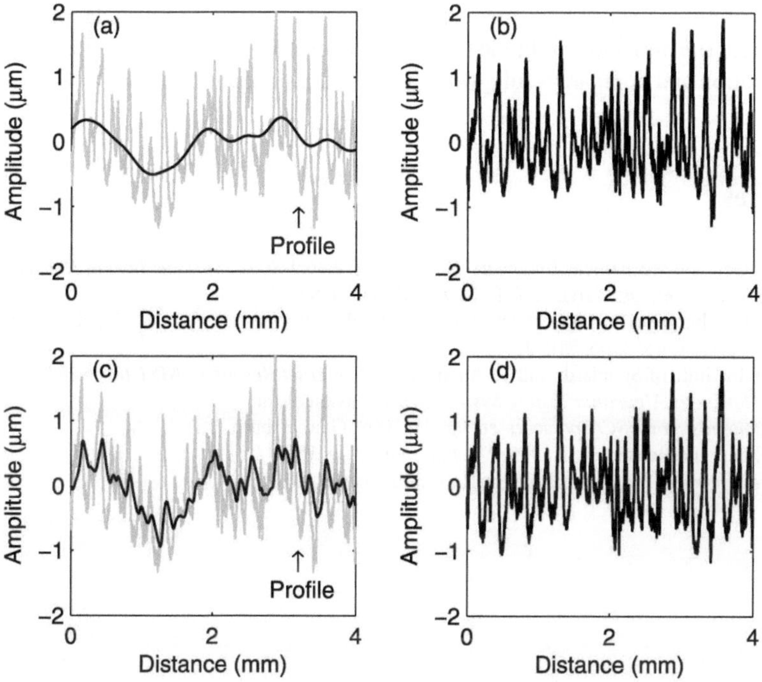

Fig. 6.4 **a** Raw profile and Gaussian low-pass filter waviness profile. **b** Roughness profile using Gaussian filter. **c** Raw profile and 2RC "mean line". **d** Roughness profile using 2RC filter. Cutoff is 0.8 mm

alleviate some of the issues with the 2RC filter. As seen in Chap. 5, the Gaussian filter is implemented digitally. It has no phase distortion effects and also the complementary definitions of the high pass and low pass allow for use of just one filter to obtain both roughness and waviness. The Gaussian filter also suffers from edge distortion effects.

Exercises

6.1. The transfer function of a 2RC low-pass filter network is given by $Hf = \frac{1}{(1+jRC\omega)^2}$, $j = \sqrt{-1}$. The low-pass filter also transmits 75% at the cutoff. Show that $RC = \frac{\lambda c}{2\pi\sqrt{3}}$, where $\omega = \frac{2\pi}{\lambda c}$. Then show that the weighting function is given by $S(x) = \frac{xA^2}{\lambda c^2} \mathrm{e}^{\frac{-xA}{\lambda c}}$, $0 \leqslant x < \infty$, $A = 2\pi\sqrt{3}$ and the transfer function is given by $Hf = \frac{1}{(1+jRC\omega)^2} = \frac{1}{\left(1+\frac{j\lambda c}{\lambda\sqrt{3}}\right)^2}$.

6.2. What is the amplitude attenuation and phase angle of a sinusoidal profile of wavelength 2 mm and amplitude 1 μm, when filtered using a 2RC high-pass filter of cutoff 0.8 mm? What is the amplitude attenuation and phase for a sinusoidal profile of wavelength 0.5 mm and amplitude 0.5 μm? Reconstruct the filtered profile using this information and compare results with those for a Gaussian filter.

References

American Society of Mechanical Engineers 2002, *ASME B46.1-2002, Surface Texture (Surface Roughness, Waviness and Lay)*, ASME. New York, NY, USA.

Bui, S.H. and Vorburger, T.V. 2007, 'Surface metrology algorithm testing system', *Precision Engineering*, vol. 31, no. 3, pp. 218–225.

The National Institute of Standards and Technology *NIST Internet Resource: NIST Internet Based Surface Metrology Algorithm Testing System Website,* available at *http://syseng.nist.gov/VSC/jsp/*, accessed Jan 25, 2008. Gaithersburg, MD, USA.

Whitehouse, D.J. and Reason, R.E. 1965, *The Equation of the Mean Line of Surface Texture Found by an Electric Wave Filter*, Rank Taylor Hobson. Leicester, UK.

Chapter 7
Filtering Roundness Profiles

7.1 Introduction

A roundness profile, such as those obtained using a rotating-spindle stationary probe roundness instrument is a closed profile, unlike open profiles obtained on planar surfaces. A nominally round part can be viewed as comprising of different surface wavelengths superimposed on it. As in the case of the surface texture analysis of planar open surfaces, these wavelengths are the fingerprint of the manufacturing processes that generated the part, and can impact function in different ways. Therefore, roundness profiles are also partitioned into different wavelength regimes, to better understand process parameters and functional performance.

There is one significant difference between open and closed surface analysis. Because a roundness instrument only captures deviations in the radial direction, it is insensitive to the actual diameter of the part. A nominally round part can be viewed as comprising of different harmonics. A harmonic is specified in terms of undulations per revolution (UPR). That is, an oval part has 2 UPR, a three-lobed part has 3 UPR *etc*. By basing our analysis on undulations per revolution instead of surface wavelengths, we no longer have to be concerned about part diameter.

The filtering process in the case of roundness profiles is therefore blind to surface wavelengths in units of length, and therefore diameter; cutoff is specified in undulations per revolution and the filter partitions a profile into different bandwidths that contain different undulations per revolution. Note that undulations per revolution is a frequency term; roundness analysis is based on frequencies while surface profile analysis is based on wavelengths.

7.2 Gaussian Filter for Roundness

The weighting function of a Gaussian low-pass filter for surface texture analysis is given by

$$S(x) = \frac{1}{\alpha\lambda c}\exp\left[-\pi\left(\frac{x}{\alpha\lambda c}\right)^2\right], \tag{7.1}$$

where $\alpha = \sqrt{\ln 2/\pi} = 0.4697$, x is the position from the origin of the weighting function, λc is the long-wavelength roughness cutoff and λ is the wavelength of different sinusoidal profiles on the surface.

In the case of cylindrical components, the cutoff is specified in undulations per revolution and is denoted by ωc. Let n be the number of points sampled on the surface, D be the diameter of the part, and Δx be the spacing in millimeters, then the distance x of the ith sampling point along the circumference is given by

$$x = i.\Delta x = i\frac{\pi D}{n}\,. \tag{7.2}$$

Also, the cutoff λc is related to the cutoff ωc by

$$\lambda c = \frac{\pi D}{\omega c}\,. \tag{7.3}$$

Substituting Eqs. 7.2 and 7.3 in Eq. 7.1, we get

$$S(i).(\pi D) = \frac{\omega c}{\alpha}\exp\left[-\pi\left(\frac{i.\omega c}{\alpha n}\right)^2\right]. \tag{7.4}$$

Equation 7.4 provides a Gaussian weighting function for a specific cutoff ωc, but it is a function of part radius (or diameter). However, because our data is reported after radius suppression, our analysis has no bearing on the radius. Therefore, we are justified in ignoring the radius term. This gives us the following expression for weighting function expressed as a function of point index i:

$$S(i) = \frac{\omega c}{\alpha}\exp\left[-\pi\left(\frac{i.\omega c}{\alpha n}\right)^2\right]. \tag{7.5}$$

The Gaussian filter for roundness is specified in the following national and international Standards: ASME B89.3.1-1972 (American Society of Mechanical Engineers 1972) and ISO/TS 12181-1, 2:2003 (International Organization for Standardization 2003a, 2003b).

Example 7.1. Plot the Gaussian weighting function for $n = 3{,}600$ points and a cutoff of 10 UPR.

The normalized Gaussian weighting function is generally generated and plotted as a function of distance in units of length. In this case, since we do not know the part radius, we therefore cannot establish the unit of length for the x axis. A cutoff, however, is also a measure of distance (in absolute terms, *i.e.*, units of length, it scales with the radius, but in roundness analysis we always suppress radius information). We first identify the number of points in each cutoff and then compute the weighting function that would span the interval from −cutoff to +cutoff. Then, we plot the weighting function with the x axis in units of distance measured as multiples of cutoff.

```
omegac = 10;                        % cutoff in units of UPR
n = 3600;                           % number of points
alpha = sqrt(log(2)/pi);
i = (-n/omegac:1:n/omegac)'; % n/omegac is the number of points
                             % in one cutoff
x = i*omegac/n;
S = (omegac/alpha).*exp(-pi*(i.*omegac/alpha/n).^2);
S = S/sum(S);
plot(x,S)
```

7.3 Amplitude Transmission Characteristics

The Fourier transform of the weighting function provides the transmission characteristics, *i.e.*, it provides the level of amplitude attenuation and phase offset of different wavelengths. The weighting function of the Gaussian filter was given in Eq. 7.5. Its Fourier transform is given by

$$Sf(\omega) = \frac{A_{\text{output}}}{A_{\text{input}}} = \exp\left(-\pi\left(\alpha\frac{\omega}{\omega c}\right)^2\right) . \tag{7.6}$$

The amplitude transmission curve is plotted in Fig. 7.1b. Note that the x-axis is in units of undulations per revolution. This plot represents a low-pass filter because low undulations per revolution correspond to long wavelengths, while high undulations per revolution correspond to short wavelengths. The transmission characteristics from Fig. 7.2 indicate that the filter transmits from 1 to almost 10 UPR with limited amplitude suppression. The filter transmits only 50% of the amplitude when the input frequency is 50 UPR, as expected. Frequencies beyond 100 UPR are almost completely suppressed.

Example 7.2. Generate the transmission characteristics of the Gaussian filter generated in Example 7.1.

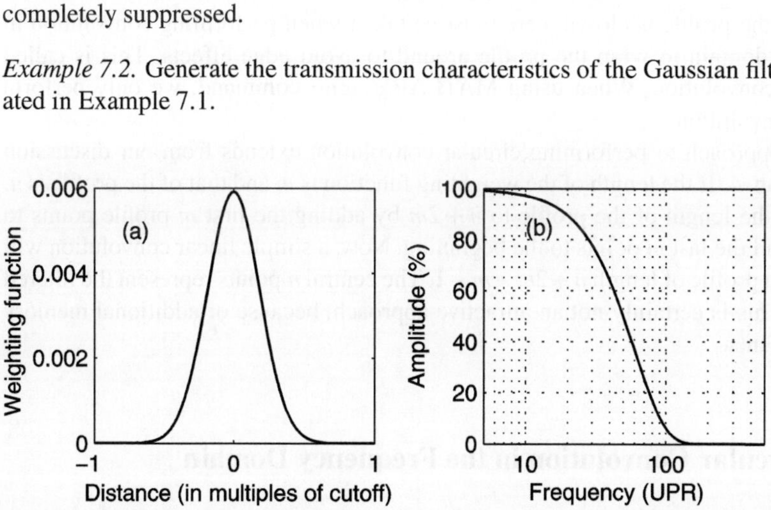

Fig. 7.1 **a** Gaussian weighting function for roundness filtering for 10 UPR cutoff. **b** Amplitude transmission characteristics of a Gaussian low-pass roundness filter of 50 UPR cutoff

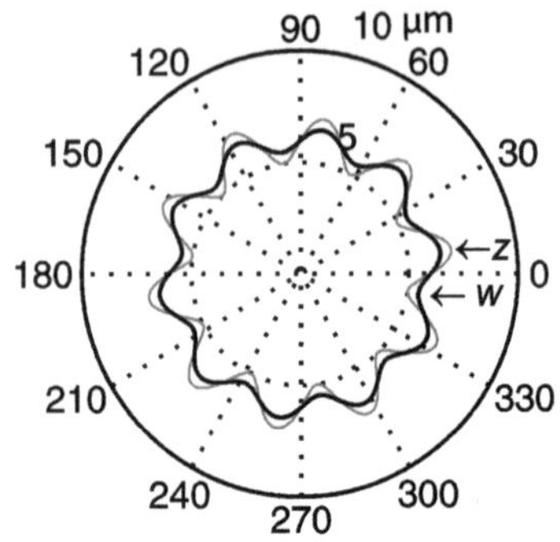

Fig. 7.2 Raw profile and filtered profile plotted on a polar plot

The following code segment generates the transmission characteristics of the filter (see Fig. 7.1b).

```
omegac = 50;                          % cutoff in UPR
upr = (5:500)';                       % generate the UPR axis
alpha = sqrt(log(2)/pi);
A = exp(-pi*(alpha*upr/omegac).^2); % From Eq. 7.6
semilogx(upr,A*100,'r');
legend(['cutoff = ' num2str(omegac)]);
xlabel('Frequency (UPR)');
ylabel('Amplitude (%)');              % plot amplitude as percent
grid on;
```

7.4 Filtering Roundness Profiles in the Time Domain

Because the profile is closed, care must be taken when performing convolution in the time domain to wrap the profile around to avoid edge effects. This is called circular convolution. When using MATLAB's `conv` command, we only perform linear convolution.

One approach to performing circular convolution extends from our discussion from Chap. 4. If the length of the weighting function is m and that of the profile is n, increase the length of the profile to $n+2m$ by adding the first m profile points to its end and the last m points to the beginning. Now, a simple linear convolution will produce a profile of length $n+2m+m-1$. The central n points represent the filtered profile. This is certainly not an attractive approach, because of additional memory requirements.

7.5 Circular Convolution in the Frequency Domain

An efficient way to perform circular convolution is using the discrete Fourier transform (DFT) in frequency domain. In general, if the DFT of profile z were multiplied by the DFT of the weighting function S, the resulting array, after inverse DFT will be equivalent to a circular convolution as discussed in Chap. 4. For filtering open profiles, we do not want to perform circular convolution and we therefore zero-pad the profile. In this case, we *do* wish to perform circular convolution therefore we do not have to zero-pad the profile. The weighting function is first generated from $x = -n/\omega c$ to $n/\omega c$. This weighting function is zero-padded to the full length to implement convolution in frequency domain. Recall from Chap. 4 that if the filter is symmetrically placed in the middle, then the filtered profile has a phase offset equal to half the profile length (extract the second half of the filtered profile and paste in the beginning to offset this).

Example 7.3. Generate sinusoidal profile R containing 3,600 points, with a 10 UPR sinusoid. Plot the radial deviations as a function of angular location.

The following code segment generates the required sinusoid.

```
n = 3600;
omega = 10;
i = (1:n)';
theta = (i-1)/n*2*pi;      % generate angle data
R = sin(2*pi/n*omega.*i); % profile with 10 UPR
% note that we use variable R to indicate radial deviation data
plot(theta,R)
```

Example 7.4. Filter the profile in Example 7.3 using the weighting function obtained in Example 7.1 and verify that the filter indeed transmits only 50% at the cutoff.

The filtering operation can be performed as shown below.

```
S1(3600,1) = 0;          % generate an array with 3600 points
S1(1441:2161,1) = S; % center filter in S1
Sf = fft(S1);            % FFT of S1
Rf = fft(R);             % FFT of R
wf = Sf.*Rf;             % term by term multiplication
w = ifft(wf);            % inverse FFT of wf
w = [w(1801:3600,1);w(1:1800,1)];
plot(theta,R,'b-',theta,w ,'r-')
```

Example 7.5. Plot profiles R and w on a polar plot of arbitrarily chosen radius that highlights the profile peaks and valleys.

The profile can be plotted on a polar plot as shown in Fig. 7.2. Polar plots may be misleading because different plot radii will cause the same radial deviation data to appear quite different. Small plot radius may even result in the profile crossing the zero (resulting in a loop at the origin).

```
polar(theta,R+6)
hold on;
polar(theta,w+6,'r-')
```

7.6 Summary

In this chapter, we have discussed some aspects to filtering closed profiles. A closed profile repeats itself, and therefore care must be taken when performing convolution. However, when filtering in the frequency domain, we realize that DFT operation does indeed result in circular convolution, and therefore we do not have to take extra care in zero-padding the profile as we did for open surfaces. Finally, because roundness instruments suppress radius information, our filter cutoffs are specified in undulations per revolution and not in units of length.

Exercises

7.1. Generate a profile R containing 3,600 points that is a sum of a 10, a 20, and a 50 UPR sinusoid, each of amplitude 1 μm. Filter this profile using a Gaussian filter of cutoff 50 UPR in the frequency domain. Plot the filtered waviness and roughness profiles. What is the value of Wa?

7.2. Perform the filtering operation in Exercise 7.1 in the time domain using convolution and verify the value of Wa.

7.3. Derive the expression for the 2RC filter equation applicable to roundness profiles from the corresponding equation for open profiles (Eq. 6.1).

7.4. Filter profile R in Exercise 7.1 using a 2RC filter of 50 UPR cutoff. Recall that the high-pass weighting function is one-sided, and produces a mean line which may be considered as an approximation to the waviness profile. Perform circular convolution as discussed in Sect. 7.4 by increasing the profile length. Discard the first and last m points (m is the length of the filter array), and then extract the first n points (see Example 6.2). Compute Wa from the mean line.

7.5. Explain the difference between the amplitudes of the filtered roughness profiles in Exercises 7.1 and 7.4.

7.6. Consider a part of diameter 10 mm measured using both a roundness instrument and a Coordinate Measuring Machine with a very small tip (300 μm diameter). Assume 3,600 points are sampled equally spaced on this part on both instruments. If a 20 UPR filter is appropriate for this part as measured on the roundness instrument, what will be the equivalent cutoff in millimeters as measured on a CMM? If the part were of diameter 5 mm instead, what would be the cutoff in millimeters?

7.7. Roundness profiles are plotted on a user-selected radius to best illustrate the undulations on a surface. In Example 7.5, we chose a radius of 6 μm. Plot profiles R and w on different radii (0 to, say, 50 μm), and compare the resulting plots. On what basis would you recommend a suitable plot radius?

References

American Society of Mechanical Engineers 1972, *ASME B89.3.1-1972, Measurement of Out-of-Roundness,* ASME. New York, NY, USA.

International Organization for Standardization 2003a, *ISO/TS 12181-1:2003, Geometrical Product Specifications (GPS) – Part 1: Vocabulary and Parameters of Roundness*, ISO. Geneva, Switzerland.

International Organization for Standardization 2003b, *ISO/TS 12181-2:2003, Geometrical Product Specifications (GPS) – Part 2: Specification Operators*, ISO. Geneva, Switzerland.

Chapter 8
Filtering 3D Surfaces

8.1 Areal Surface Texture Analysis

Areal surface texture analysis has increased in popularity in recent years, due to the availability of a variety of optical microscopy techniques coupled with increasing computing capability. The recent trend towards 3D analysis has also resulted in both ASME B46.1 (American Society of Mechanical Engineers 2002) and ISO/CD TS 25178-2:2005 (International Organization for Standardization 2005) incorporating these methods. Three-dimensional surface texture analysis is described in several recent publications; two comprehensive texts include Stout and Blunt (2000) and Blunt and Jiang (2003).

The techniques involved in data processing such as filtering and parameter computation are an extension of the methods previously discussed. We discuss the implementation of the Gaussian filter in this chapter because it is the most widely used filter today. It can be implemented quite easily using 2D convolution.

8.2 2D Convolution

Consider an example surface containing four rows and four columns of data as shown in Fig. 8.1. The 2D convolution of this surface with a rectangular filter of size 3×3 is illustrated in Fig. 8.1. The filter is positioned at the top-left of the surface, so that it overlaps just one element of the surface. The weighted sum is calculated at that point. The filter is then stepped to the right by one unit and the weighted sum is computed again. This process is repeated until the filter just overlaps the rightmost element of the first row of the surface. The filter is then moved back all the way to the left and stepped down by one unit and a weighted sum is computed. If the surface array is of size $ny \times nx$ and the filter is of size $my \times mx$, the resulting 2D convolution array is of size $ny + my - 1 \times nx + mx - 1$. Note that the filter is generally normalized to unit sum prior to convolution.

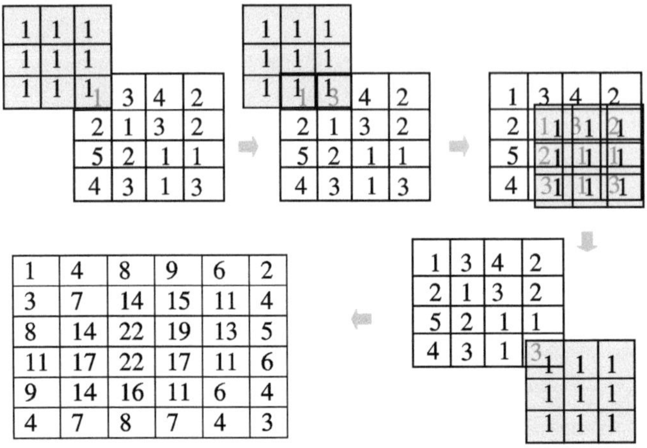

Fig. 8.1 Convolution of an area map using a rectangular filter

8.3 Gaussian Filter in 3D

The weighting function of the 3D Gaussian filter is the product of two Gaussian filters:

$$S(x,y) = \frac{1}{\alpha^2.\lambda xc.\lambda yc} \exp\left[-\left[\pi \left(\frac{x}{\alpha.\lambda xc} \right)^2 + \pi \left(\frac{y}{\alpha.\lambda yc} \right) \right]^2 \right], \qquad (8.1)$$

where λxc and λyc are the cutoffs in x and y, and $\alpha = \sqrt{\ln 2/\pi} = 0.4697$. This definition is simply a 3D extension of the 2D filter definition in Chap. 5. The filter can be implemented using 2D convolution in the spatial domain or using the Fourier transform multiplication method in the frequency domain. The latter is computationally more efficient, but the former is more easily implemented as shown in the examples below.

8.4 A Note on Indices

Let a 2D grid of points represent a digitized surface topography map (or a weighting function). Let any row extracted from this 2D grid represent a profile along the x axis, and any column extracted from this grid represent a profile along the y axis. Let i be an index for points along the x axis and j be an index along the y axis. Then, the matrix that stores this 2D grid in MATLAB is given by $z(j,i)$, not $z(i,j)$. This is because 2D matrices are generally stored and accessed in (*row*, *column*) format where *row* represents an index along the y axis and *column* represents an index along the x axis. Figure 8.2 illustrates this.

→ $x(i)$, columns

↓ $y(j)$ rows

y\x	1	2	3	4	5	6
1	1	4	8	9	6	2
2	3	7	14	15	11	4
3	8	14	22	19	13	5
4	11	17	22	17	11	6
5	9	14	16	11	6	4
6	4	7	8	7	4	3

Fig. 8.2 Let the data shown represent a surface topography map containing six points along each direction. Let the spacing between points be unity. Let the function $f(x,y)$ represent the surface. Then, $f(x,y)$ at $(5,3)$ is $f(5,3) = 13$. Let the discrete data be stored in a matrix z. Then, $z(3,5)$ [not $z(5,3)$] $= 13$

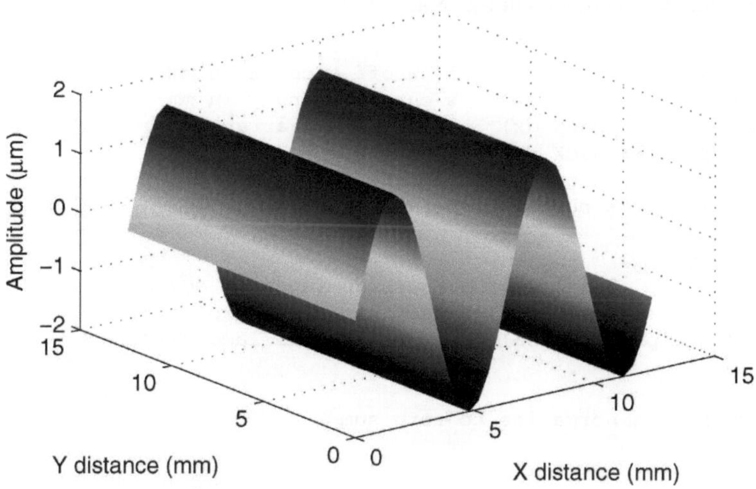

Fig. 8.3 Sinusoidal surface z

Example 8.1. Generate a 3D sinusoidal surface $z = 2\sin(2\pi x/6.4), 0 \leqslant x < 12.8$ mm, $0 \leqslant y < 12.8$ mm, using a sampling interval of 100 µm along both directions.

The unidirectional sinusoidal surface z can be generated as shown below. It is plotted in Fig. 8.3.

```
dx = 0.1;               % spacing in x in mm
dy = 0.1;               % spacing in y in mm
nx = 128;               % number of points along x
ny = 128;               % number of points along y
x = (0:1:nx-1)*dx;  % generate x array
y = (0:1:ny-1)*dy;  % generate y array
for i = 1:nx
    for j = 1:ny    % notice that i is the index of the x array,
                    % while j is the index of the y array. The
                    % columns of z must contain the sinusoid if it
                    % is to be oriented along the X axis
```

```
        z(j,i) = 2*sin(2*pi*x(i)/6.4);
    end
end
z = z - mean(mean(z)); % shift to zero mean
surf(x,y,z);
xlabel('X distance (mm)');
ylabel('Y distance (mm)');
zlabel('Amplitude (\mum)');
shading interp;
```

Example 8.2. Generate a Gaussian filter with $\lambda xc = \lambda yc = 6.4$ mm, using a sampling interval of 100 μm along both directions.

The Gaussian filter S can be generated in the interval $-\lambda xc \leqslant x \leqslant \lambda xc$, $-\lambda yc \leqslant y \leqslant \lambda yc$ as shown below. It is plotted in Fig. 8.4.

```
alpha = sqrt(log(2)/pi);
lamdacX = 6.4;                    % cutoff along x in mm
lamdacY = 6.4;                    % cutoff along y in mm
x = (-lamdacX:dx:lamdacX-dx)'; % generate x array
y = (-lamdacY:dy:lamdacY-dy)'; % generate y array
mx = size(x,1);     % number of points along x
my = size(y,1);     % number of points along y
for i = 1:mx
    for j=1:my
                    % sample the Gaussian function equation
        S(j,i) = (1/(alpha^2*lamdacX*lamdacY))*exp(-pi*
        (x(i)/alpha/lamdacX)^2-pi*(y(j)/alpha/lamdacY)^2);
    end
end
S=S/sum(sum(S)); % normalize to zero sum
surf (x,y,S);
```

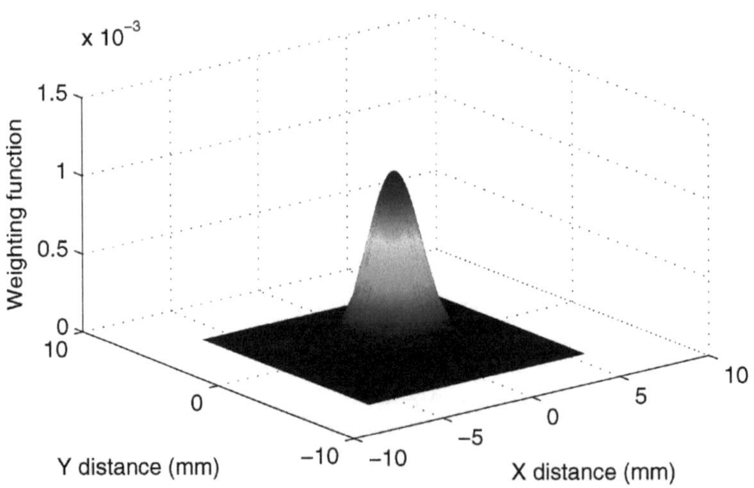

Fig. 8.4 Gaussian filter with $\lambda xc = \lambda yc = 6.4$ mm

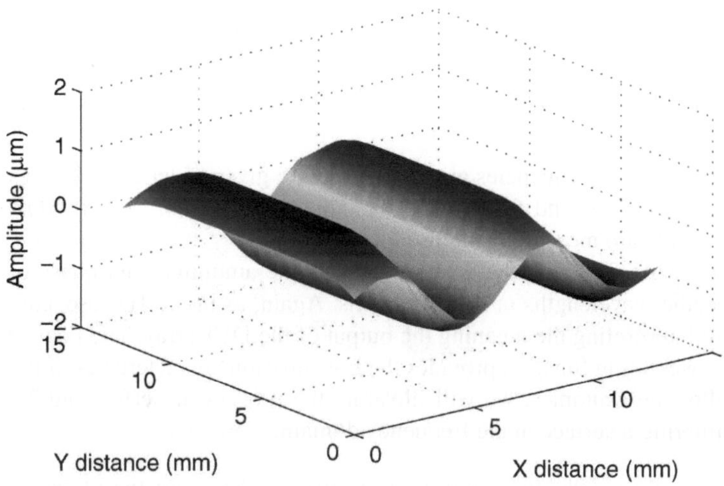

Fig. 8.5 Filtered waviness surface

```
xlabel('X distance (mm)');
ylabel('Y distance (mm)');
zlabel('Weighting function');
```

Example 8.3. Filter surface *z* in Example 8.1 using the filter *S* in Example 8.2. Plot the waviness surface and determine the amplitude attenuation.

We use MATLAB's inbuilt 2D convolution function `conv2`. The amplitude of the filtered surface can be determined from the plot in Fig. 8.5 as 1 µm. Therefore, the amplitude attenuation is 50%, as expected. For clarity, a single profile can be extracted and plotted along with the original profile. We leave it as an exercise for the reader to perform this.

```
x = (0:1:nx-1)'*dx;                  % generate x array
y = (0:1:ny-1)'*dy;                  % generate y array
C = conv2(z,S);                      % 2D convolution
w = C(my/2+1:ny+my/2,mx/2+1:nx+mx/2); % extract central portions
                                     % of the surface
surf(x,y,w)
xlabel('X distance (mm)');
ylabel('Y distance (mm)');
zlabel('Amplitude (\mum)');
```

8.5 Frequency-Domain Filtering in 3D

Three-dimensional surfaces can be filtered in the frequency domain in a manner very similar to that discussed in Chap. 3. The Fourier transform of the surface is multiplied by the Fourier transform of the weighting function An inverse transform then provides the filtered surface. For this purpose, we use the 2D DFT defined as follows:

$$zf(f_y, f_x) = \sum_{k=1}^{ny} \sum_{j=1}^{nx} z(k,j)\,\mathrm{e}^{-i2\pi[(j-1)(fx)\Delta x+(k-1)(fy)\Delta y]}\,, \tag{8.2}$$

where fx and fy (surface frequencies along x and y) are given in terms of indices nx and ny as $fx = (j-1)/lx$ and $fy = (k-1)/ly$, j and k range from 1 to nx and ny respectively; lx and ly are measured lengths along x and y.

When working with 2D DFT, the array represents the amplitude and phase of sinusoids of specific wavelengths in two directions. Again, as in the 1D case, care must be taken in interpreting the meaning the output of the DFT array. The Fourier transform of the weighting function provides the transmission characteristics of the filter. In the following examples, we will illustrate the process of performing 2D DFT and then filtering a surface in the frequency domain.

Example 8.4. Generate and plot the amplitude transmission characteristics of the 2D Gaussian filter.

The 2D DFT of the weighting function defined in Eq. 8.1 provides the amplitude transmission characteristics. In MATLAB, we use the `fft2` function. Array Sf is of the same size as S. Care must be taken to interpret the x and y axes in the plot. Without special effort, the axes are only indices of points in the Sf array. From Chap. 3, we can interpret the axes as follows. The first row of Sf is the result of a 1D DFT along the x direction; that is, the first term is the DC component. Subsequent terms represent wavelengths in decreasing order until the middle (the Nyquist frequency). The second half represents wavelengths in increasing order. The first column of Sf can be interpreted as the result of a 1D DFT along the y direction. For purposes of plotting, we add the wavelength information as superimposed text on the axes. The plot is shown in Fig. 8.6.

```
Sf = fft2(S);
mesh(abs(Sf));
zlabel('Scaled amplitude (\mum)');
```

However, for plotting purposes, it is sometimes desirable to have the DC term in the middle with increasing frequencies in x to the right, and increasing frequencies in y to the top. This is achieved using the function `fftshift` in MATLAB, see Fig. 8.7.

```
Sf_shift = fftshift(Sf);
mesh(abs(Sf_shift));
```

Example 8.5. Filter surface z in Example 8.1 in the frequency domain using the Gaussian filter.

The following code segment illustrates the filtering process. It is not required that Sf and zf be shifted as shown below. However, should we decide to view the Fourier transform, then the shifted transform is easier to comprehend; this is especially true for the transform of sinusoidal surfaces.

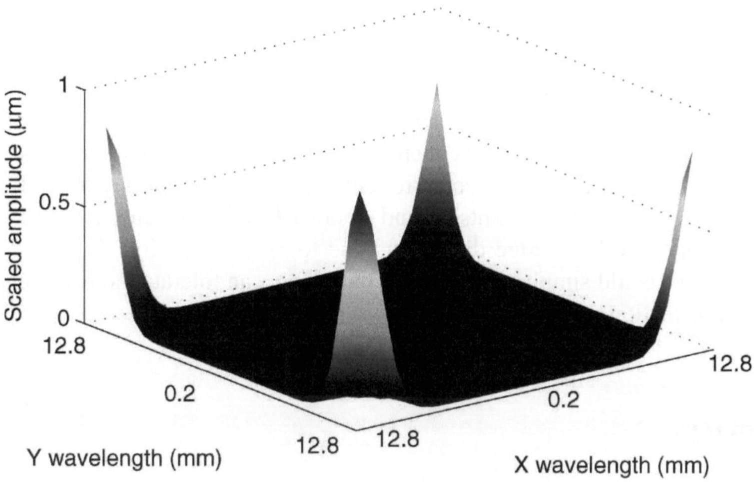

Fig. 8.6 2D Fast Fourier transform (FFT) of the weighting function *S* (note that the *x* and *y* axes legends are superimposed text representing surface wavelengths)

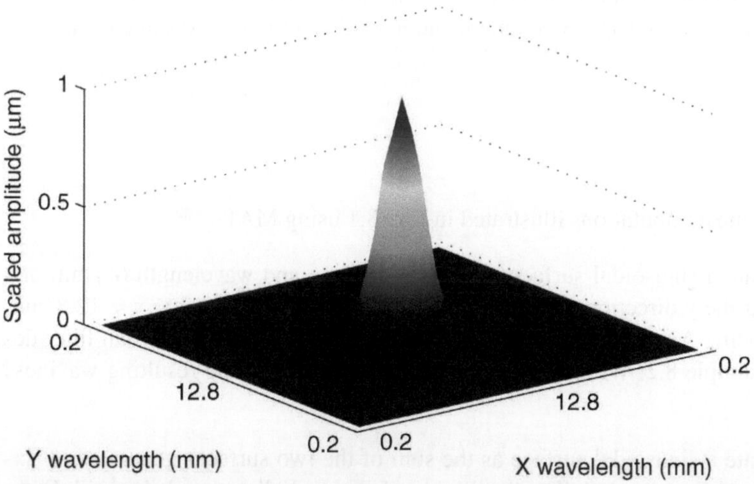

Fig. 8.7 2D FFT of the weighting function *S* after re-shifting the origin (note that the *x* and *y* axes legends are superimposed text representing surface wavelengths)

```
Sf = fftshift(fft2(S));      % compute 2D fft of S and shift origin
zf = fftshift(fft2(z));      % compute 2D fft of z and shift origin
wf = Sf.*zf;                 % term by term multiplication
w = ifft2(ifftshift(wf));    % shift the origin back and take the
                             % inverse FFT
w = [w(65:128,:); w(1:64,:)]; % see Example 4.6 for why we juggle
                                % the quadrants around
w = [w(:,65:128) w(:,1:64)];
mesh(x,y,z); hold on; mesh(x,y,real(w))
```

```
xlabel('X distance (mm)');
ylabel('Y distance (mm)');
zlabel('Amplitude (\mum)');
```

It should be noted that by not zero-padding the surface in both directions, we have performed circular convolution. We have therefore two problems to overcome. First, we have a phase offset in the output equal to half the length in each direction. To overcome this, we juggle the quadrants around (follows from our discussion in Example 4.6). Second, we have edge distortions due to assumed periodicity in the profile. For this, we could simply discard edge data if we can tolerate the resulting loss of some information.

8.6 Summary

3D area maps are increasingly used today in place of 2D profiles. Filtering 3D surfaces is a key aspect to 3D surface texture analysis. In this chapter, we have discussed the Gaussian filter as applied to 3D surface texture analysis. We have built on our discussion in Chaps. 3, 4, and 5, and extended it to the next dimension.

Exercises

8.1. Verify the computations illustrated in Fig. 8.1 using MATLAB.

8.2. Generate a sinusoidal surface of amplitude 2 μm and wavelength 6.4 mm oriented along the y direction. Generate the surface in the interval $0 \leqslant x < 12.8$ mm, using a spacing $\Delta x = \Delta y = 0.1$ mm. Filter the surface using the Gaussian filter defined in Example 8.2. What is the amplitude attenuation of the resulting waviness surface?

8.3. Generate a sinusoidal surface as the sum of the two surfaces generated in Example 8.1 and Exercise 8.2. The resulting surface is a bidirectional sinusoid. Filter this surface using the Gaussian filter defined in Example 8.2 and plot the resulting waviness surface.

8.4. The surface generated in Exercise 8.3 has two sinusoidal components, one oriented along x and the other along y, both of wavelength 6.4 mm. Suggest suitable cutoffs for a Gaussian low-pass filter along x and y so that the sinusoid along x will be retained in the waviness surface, while that along y will be suppressed. Verify your answer using MATLAB. (Note: there can be several answers; justify your choice).

8.5. For a matrix A defined as $A = [1\ 3\ 2;\ 2\ 4\ 1;\ 0\ 3\ 2]$, compute the 2D DFT using Eq. 8.2, and verify the result using MATLAB.

8.6. In Example 8.5, we noted that the result was actually a circular convolution. Zero-pad the surface and the filter along both directions (with the surface placed in the middle), so that the new array is of size 384 × 384. Filter the surface in the frequency domain and compare the result with that obtained by convolution. Are the results identical at the edges?

References

American Society of Mechanical Engineers 2002, *ASME B46.1-2002, Surface Texture (Surface Roughness, Waviness and Lay)*, ASME. New York, NY, USA.

Blunt, L. and Jiang, X. (eds) 2003, *Advanced Techniques for Assessment Surface Topography: Development of a Basis for 3D Surface Texture Standards*, Kogan Page Science. London, UK.

International Organization for Standardization 2005, *ISO/CD TS 25178-2:2005, Geometrical Product Specifications (GPS) – Surface Texture: Areal – Part 2: Terms, Definitions And Surface Texture Parameters*, ISO. Geneva, Switzerland.

Stout, K. and Blunt, L. 2000, *Three Dimensional Surface Topography*, 2nd edn, Butterworth-Heinemann.

Part II
Advanced Filtering

We describe several advanced filtering techniques in Part II. These filters are designed to specifically overcome some of the limitations of the 2RC and Gaussian filters. In Chap. 9, we describe the Gaussian regression filter that was proposed to overcome the edge effect problem associated with the Gaussian filter. In Chap. 10, we discuss the proposed spline filter that was designed for the same purpose. Different methods for filtering in presence of outliers are the focus of Chap. 11. We devote Chap. 12 to discuss an entirely different class of filters, namely, morphological filters that are based on rolling ball/sphere or sliding line/plane envelopes. In Chap. 13, we briefly describe the evolution of both the rolling ball filter and the traditional Gaussian-type filter into multi-scale filtering to achieve fine partitioning of the surface.

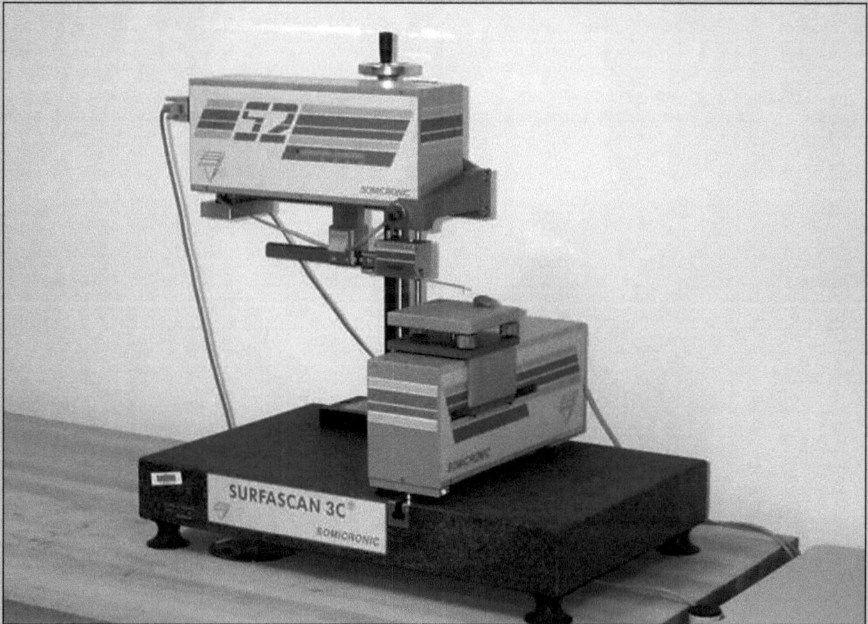

Fig. II A stylus-based surface texture measurement instrument (The Somicronic Surfascan 3C at the Center for Precision Metrology at UNC Charlotte)

Chapter 9
Gaussian Regression Filters

9.1 Introduction

As mentioned in Chap. 2, the Gaussian filter suffers from some limitations such as edge distortion, inability to deal with large form, and outlier-induced distortion. Several new filters have been proposed recently that overcome these limitations. One such is the Gaussian regression filter. Proposed by Brinkmann *et al.* (2000, 2001), it is a fairly recent development in surface texture analysis and is currently a draft international Standard. It is based on the Gaussian filter, but combines the traditional Gaussian filter with either a zero-order or a second-order polynomial fit to reduce edge effect and large form issues.

In the case of a traditional Gaussian filter, the waviness profile element at any location is evaluated as the weighted average of the profile and the Gaussian filter whose central element is placed at that location. In the case of the Gaussian regression filter, the waviness profile element at any location is obtained by minimizing a function E, which is the Gaussian weighted squared difference of the profile and a polynomial of either zero or second order. The coefficients of the polynomial are determined by minimizing E. The polynomial evaluated at that location is itself the value of the waviness profile element at that location.

If the polynomial were of zero order (a constant term only), the filter behaves identically with the traditional Gaussian filter, except at the edges, where it shows no distortion. If the polynomial were of second order, the filter shows no edge distortion and also performs well in the presence of large form. A first-order filter is also feasible (Seewig 2005). We discuss only the zero and second order here.

Equations in this chapter are reproduced from Brinkmann *et al.* (2000) with permission. The Gaussian regression filter is defined by the following general regression arrangement:

$$\text{Minimize} \quad E(k) = \sum_{p=1}^{n} (z(p) - w(k))^2 S(k,p)\Delta x\,, \tag{9.1}$$

where
z is the profile element,
w, the mean line element,
n, the number of points,
k, the index for the location of the center of the weighting function,
p, the index for profile points,
Δx, the spacing,
λc, the cutoff, and
S, the weighting function as given by

$$S(k,p) = \frac{1}{\lambda c\sqrt{\ln(2)}}\exp(-\frac{\pi^2}{\ln(2)}\frac{((k-p)\Delta x)^2}{\lambda c^2})\,. \tag{9.2}$$

In the regression approach, the weighting function is calculated for every point on the profile and the objective function is minimized to compute the element of the mean line w. A zero-order regression filter minimizes the square of the deviation of the profile element from a horizontal line at each point. Thus, the minimization problem involves computing a constant term at every point of the profile. The second-order regression filter minimizes the square of the deviation of the profile element from a quadratic curve at each point. Here, the minimization involves finding three coefficients of the quadratic curve at every point of the profile.

9.2 Zero-Order Gaussian Regression Filter

In the case of the zero-order Gaussian regression filter, $w(k)$ is simply a constant term $C(k)$. The problem is to determine this constant for every point in the profile by minimizing $E(k)$, $1 \leq k \leq n$, where

$$E(k) = \sum_{p=1}^{n} (z(p) - C(k))^2 S(k,p)\Delta x\,. \tag{9.3}$$

Taking partial derivatives with respect to C,

$$\frac{\partial E}{\partial C} = \frac{\partial}{\partial C}\sum_{p=1}^{n}(z(p)-C(k))^2 S(k,p)\Delta x = 2\sum_{p=1}^{n}(z(p)-C(k))S(k,p)\Delta x = 0\,. \tag{9.4}$$

Therefore,

$$C(k) = \frac{\sum_{p=1}^{n} z(p)S(k,p)\Delta x}{\sum_{p=1}^{n} S(k,p)\Delta x}\,. \tag{9.5}$$

Thus, the zero-order regression filter is equivalent to the convolution given below:

$$C(k) = \sum_{p=1}^{n} z(p)S_{MOD}(k,p)\,, \tag{9.6}$$

where the modified weighting function is given by

$$S_{MOD}(k,p) = \frac{S(k,p)}{\sum\limits_{p=1}^{n} S(k,p)} . \tag{9.7}$$

It can be seen that the weighting function of this filter is modified in the first and last cutoff sections as shown in Example 9.1. However, it remains unchanged in the interior regions of the profile.

Example 9.1. Generate the modified weighting function of the zero-order Gaussian regression filter. Assume $\lambda c = 0.8$ mm, profile length $l = 8$ mm, and sampling interval $\Delta x = 1$ μm.

The weighting function of the zero-order Gaussian regression filter is generated over the entire profile length, and is modified near the edges as defined in Eq. 9.7. Figure 9.1 shows a plot of the weighting function at ten different locations along the profile. Only the first 2 mm of the weighting function is plotted, to clearly show the changing shape near the edges. The following MATLAB code plots the modified Gaussian weighting function.

```
n = 8000;              % number of profile points
const = sqrt(log(2)/2/pi);
lambdac = 0.8;         % lambdac in mm
dx = 0.001;            % spacing in mm
x = (0:1:n-1)'*dx;     % generate x axis data from 0 to 8 mm
for k=1:50:500
   p=(1:1:n)';         % For each position k (center of the filter),
                       % generate the filter over the entire profile
                       % length
   S = (1/sqrt(2*pi)/const/lambdac).*exp(-0.5*((k-
   p)*dx/const/lambdac).^2);% generate weighting function
   SMOD = S/sum(S);    % modified weighting function
   plot(x(1:2000),SMOD(1:2000));
   hold on;
end
```

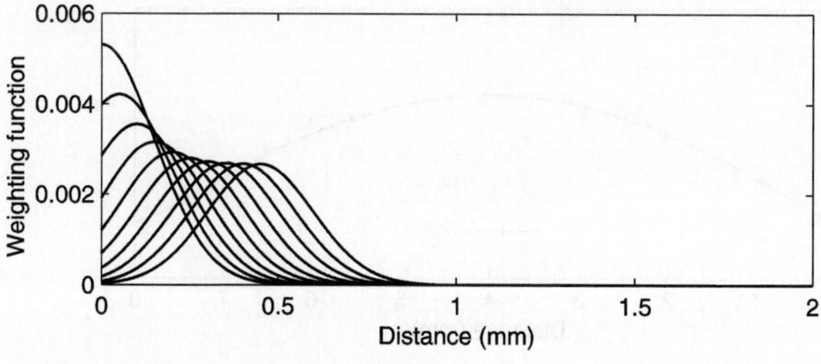

Fig. 9.1 Modified Gaussian weighting function

```
xlabel('Distance (mm)');
ylabel('Weighting function');
```

Example 9.2. Generate a sinusoidal profile z of amplitude 2 μm and wavelength 16 mm, in the interval $0 \leq x < 8$ mm, sampled at a spacing $\Delta x = 1$ μm. Superimpose normally distributed random noise of zero mean and 0.1 μm standard deviation. Remove any residual slope in the data. Filter this profile using (a) a Gaussian filter of 0.8 mm cutoff, and (b) a zero-order Gaussian regression filter of 0.8 mm cutoff. Notice the rather large edge effect in the Gaussian filter waviness profile and the absence of this effect in the Gaussian regression waviness profile. In situations where edge effect cannot be tolerated, such as in measurement of a short trace, the regression filter is useful.

The following MATLAB code generates profile z and the waviness profiles using a Gaussian and zero-order Gaussian regression filter. Notice the reduced edge distortion of the Gaussian regression mean line in Fig. 9.2. We use the MATLAB function `normrnd` from the statistics toolbox to generate normally distributed random noise. The MATLAB function `rand` can be used to generate uniform noise in the interval $[0, 1]$, and `randn` can be used to generate normally distributed noise of zero mean and unit standard deviation. In the case of the zero-order Gaussian regression filter, we do not use MATLAB's `conv` function for filtering. This is because we generate the filter over the entire length of the profile for each profile location (the filter looks different near the edges as shown in Fig. 9.1) and perform term by term multiplication as shown below.

```
n = 8000;                    % number of profile points
dx = 0.001;                  % spacing in mm
x = (0:1:n-1)'*dx;           % generate x axis data from 0 to 8 mm
z = detrend(2*sin(2*pi.*x/16)+normrnd(0,0.1*ones(n,1)));
                             % generate profile z using normrnd
                             % function
                             % FILTER USING GAUSSIAN
lambdac = 0.8;               % lambdac in mm
xg = (-lambdac:dx:lambdac)';  % x axis for generating filter
```

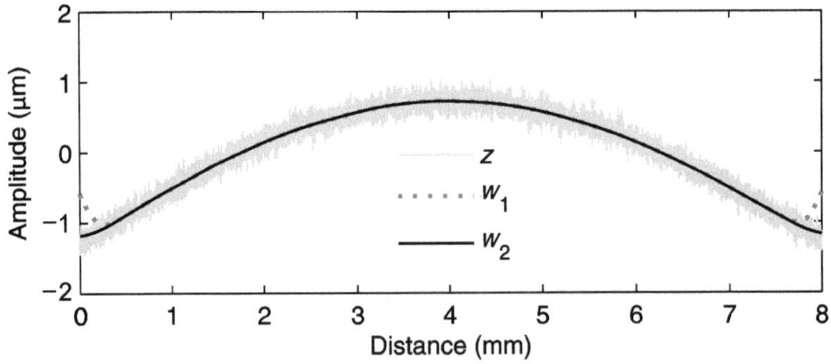

Fig. 9.2 Profile z and waviness profiles w_1, w_2

```
                             % -cutoff to +cutoff
alpha = 0.4697;
S = (1/(alpha*lambdac)).*exp(-pi*(xg/(alpha*lambdac)).^2);
                             % Gaussian filter
S = S/sum(S);                % normalize filter sum to unity
w1 = conv(S,z);              % convolution of filter and profile
w1 = w1(801:8800,1);         % extract central 8000 points of profile
                             % ZERO ORDER GAUSSIAN REGRESSION
const = sqrt(log(2)/2/pi/pi);
for k=1:n
   p=(1:1:n)';               % For each position k (center of the
                             % filter), generate the filter over the
                             % entire profile length
   S = (1/sqrt(2*pi)/const/lambdac).*exp(-0.5*((k-
p)*dx/const/lambdac).^2);   % generate weighting function
   SMOD = S/sum(S);          % normalize filter to unit sum
   w2(k,1) = sum(SMOD.*z);   % sum of products (this step replaces
                             % the convolution)
end
plot(x,z,'g',x,w1,'r',x,w2,'k');
xlabel('Distance (mm)');
ylabel('Amplitude (\mum)');
```

9.3 Second-Order Gaussian Regression Filter

The second-order Gaussian regression filter is realized by minimizing the deviations between the input data set and a second-order polynomial to obtain the element of the mean line at each point of the profile.

The implementation of the second-order filter is as follows. For each profile point k, the weighting function S is generated according to Eq. 9.2. We then determine the terms of the quadratic $w(k,p) = Ax(k,p)^2 + Bx(k,p) + C$, where $x(k,p) = (k-p)\Delta x$, $p = 1:n$, by minimizing

$$
\begin{aligned}
E(k) &= \sum_{p=1}^{n} (z(p) - w(k))^2 S(k,p)\Delta x \\
&= \sum_{p=1}^{n} (z(p) - Ax(k,p)^2 - Bx(k,p) - C)^2 S(k,p)\Delta x\,. \qquad (9.8)
\end{aligned}
$$

Note carefully that in the above formulation, w is an array containing n points. It represents a quadratic that can be described in terms of x and three parameters A, B, and C, whose values change for every point k. We have chosen to describe the x coordinate such that it is always zero for the kth point which is the center of the weighting function. Therefore, x ranges from $(k-1)\Delta x$ to $(k-n)\Delta x$ instead of 0 to $(n-1)\Delta x$. The advantage in doing so is that the value of the waviness profile at the kth location is given by $p = k$, that is, $w(k,k) = Ax(k,k)^2 + Bx(k,k) + C$. Since $x(k,k) = 0$, $w(k,k) = C$.

Taking partial derivatives with respect to A, B, and C and equating to zero, we obtain the following system of three equations and three variables which can be easily solved to obtain A, B, and C:

$$\sum_{p=1}^{n}(z(p)-Ax(k,p)^2-Bx(k,p)-C)S(k,p)x(k,p)^2\Delta x=0$$
$$\sum_{p=1}^{n}(z(p)-Ax(k,p)^2-Bx(k,p)-C)S(k,p)x(k,p)\Delta x=0\,. \tag{9.9}$$
$$\sum_{p=1}^{n}(z(p)-Ax(k,p)^2-Bx(k,p)-C)S(k,p)\Delta x=0$$

Example 9.3. Filter profile z in Example 9.2 using the second-order Gaussian regression filter of 0.8 mm cutoff. Compare the waviness profile with that obtained using the zero-order Gaussian regression filter.

The following MATLAB code generates the second-order Gaussian regression waviness profile w_3. Notice that w_3 has superior edge performance. However, it is computationally more intense and its performance in the presence of large form is only marginally superior (profile w_3 is less influenced by the overall curvature and is not weighted down; it hugs the texture better and is lifted above profile w_2. This can be seen by zooming into the central portion of the profile). Profiles z, w_2, and w_3 are shown in Fig. 9.3.

```
const = sqrt(log(2)/2/pi/pi);
for k = 1:n
   p = (1:1:n)';   % For each position k (center of the filter),
                   % generate the filter over the entire profile
                   % length
   S = (1/sqrt(2*pi*pi)/const/lambdac).*exp(-0.5*((k-
p)*dx/const/lambdac).^2);% generate weighting function
   S = S/sum(S);   % normalize to unit sum
% Proceed with regression - solve the system of three equations
```

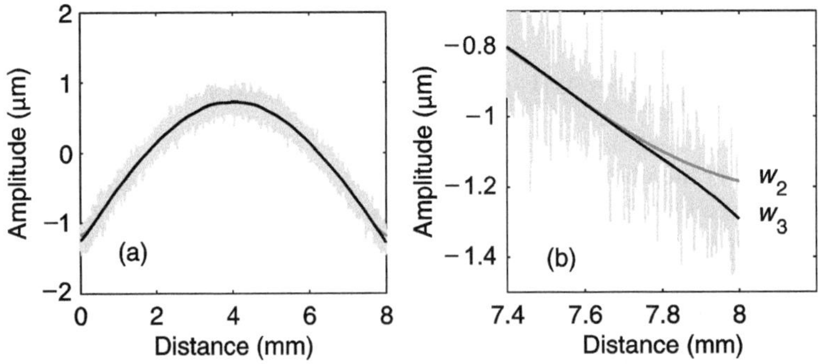

Fig. 9.3 **a** Profile z and waviness profiles w_2 and w_3 over the entire 8 mm length and **b** the edge effect shown in the last 0.6 mm

```
% and three variables given in Eq. 9.9. Ignore the dx term in
%  Eq. 9.9 because it is a constant.
   x= (k-p)*dx;
   M(1,1)  = sum(S.*x.^4);
   M(1,2)  = sum(S.*x.^3);
   M(1,3)  = sum(S.*x.^2);
   M(2,1)  = M(1,2);
   M(2,2)  = M(1,3);
   M(2,3)  = sum(S.*x);
   M(3,1)  = M(1,3);
   M(3,2)  = M(2,3);
   M(3,3)  = sum(S);
   Q(1,1)  = sum(z.*S.*x.^2);
   Q(2,1)  = sum(z.*S.*x);
   Q(3,1)  = sum(z.*S);
   P = inv(M) * Q;  % determine array P containing the values of
                   % three unknown
   A = P(1); B = P(2); C =  P(3);
   w3(k,1) = C;    % determine waviness profile value at that
                   % location
end
plot(x,z,'g',x,w2,'r',x,w3,'k');
xlabel('Distance (mm)');
ylabel('Amplitude (\mum)');
```

9.4 3D Zero-Order Gaussian Regression Filter

The zero-order regression filter in 3D is equivalent to the convolution given below:

$$C(kx,ky) = \sum_{py=1}^{ny} \sum_{px=1}^{nx} z(lx,ly) S_{MOD}(kx,px,ky,py)\,, \tag{9.10}$$

where the modified weighting function is given by

$$S_{MOD}(kx,px,ky,py) = \frac{S(kx,px,ky,py)}{\sum\limits_{ly=1}^{ny} \sum\limits_{lx=1}^{nx} S(kx,px,ky,py)}\,, \tag{9.11}$$

and S is given by

$$S(kx,px,ky,py) = \frac{1}{\alpha^2 \lambda xc \lambda yc} \exp\left[-\left[\pi\left(\frac{px-kx}{\alpha\lambda xc}\right)^2 + \pi\left(\frac{py-ky}{\alpha\lambda yc}\right)^2\right]\right]. \tag{9.12}$$

Notice that we have simply extended Eqs. 9.2, 9.6, and 9.7 into another dimension. In Eqs. 9.10, 9.11, and 9.12, the following terminology holds: z is the profile element, C, the mean line, nx and ny, the number of points along x and y, kx and ky, the index for the location of the weighting function along x and y, and px and py the index for profile points along x and y.

Example 9.4. Generate the modified weighting function of the 3D zero-order Gaussian regression filter. Assume $\lambda xc = 0.016$ mm and $\lambda yc = 0.128$ mm, profile length $lx = ly = 0.128$ mm, and sampling interval $\Delta x = \Delta y = 1$ μm.

The following demonstrates the generation of the 3D zero-order Gaussian regression filter at different points on the surface.

```
nx = 128;                    % number of profile points along x
ny = 128;                    % number of profile points along y
const = sqrt(log(2)/2/pi/pi);
lambdax = 0.016;             % lambdac in mm
lambday = 0.128;
dx = 0.001;                  % spacing in mm
dy = 0.001;
for kx = 1:16:nx
    for ky = 1:16:ny
        px = 1:1:nx;
        py = 1:1:ny;
        S = (1/sqrt(2*pi)/const^2/lambdax/lambday)*[[exp(-0.5*
((py-ky)*dy/const/lambday).^2)]'*[exp(-0.5*((px-kx)
*dx/const/lambdax).^2)]];
        S = S/sum(sum(S));
        mesh(S); xlabel('X distance (mm)'); ylabel('X distance
(mm)');
        pause;
        clf;
    end
end
```

Note that we still follow the convention described in Sect. 8.4. As we step through the sequence of plots, it can be seen that the cutoff along y is indeed large.

9.5 3D Second-Order Gaussian Regression Filter

The 3D second-order Gaussian regression filter is realized by minimizing (E_{3D}) the deviations between the input data set and a second-order polynomial surface to obtain the element of the mean line at each point of the surface. For each profile point (kx, ky), the weighting function S is generated according to Eq. 9.12. The terms of the quadratic

$$w(kx,px,ky,py) = Ax(kx,px)^2 + By(ky,py)^2 + Cx(kx,px) + Dy(ky,py) + \\ + Ex(kx,px)y(ky,py) + F \tag{9.13}$$

$$x(kx,px) = (kx-px)\Delta x\,, \quad px = 1:nx\,, \quad y(ky,py) = (ky-py)\Delta y\,, \quad py = 1:ny \tag{9.14}$$

are found by minimizing

$$E_{3D}(kx,ky) = \sum_{py=1}^{ny} \sum_{px=1}^{nx} (z(px,py) - w(kx,px,ky,py))^2 S(kx,px,ky,py) . \quad (9.15)$$

We leave it as an exercise to the reader to determine the unknown parameters by partial differentiation in a manner similar to the 2D case.

9.6 Summary

The Gaussian regression filter overcomes the two problems of the Gaussian filter – problem of edge distortion and inability to deal with large form. The zero-order Gaussian regression filter is an efficient modification of the Gaussian filter to overcome edge distortion. The second-order filter is computationally more intense, but its performance in the presence of large form is superior to the zero-order filter. These filters, however, suffer from another issue. In the presence of outliers, the mean lines generated using these filters tend to be distorted. That is, they are not robust against outliers. In Chap. 11, we discuss a modification of the regression filter that is robust against outliers, but in the meantime, we focus our attention on yet another filter, the spline filter that also addresses the two issues of edge distortion and large form problem. The spline filter is fast gaining ground as the filter of choice because of its steeper transmission characteristics and reduced computational burden, despite a fairly major limitation, the lack of an expression for a weighting function. We discuss the spline filter next, in Chap. 10.

Exercises

9.1. Generate a Gaussian filter, using Eq. 9.2, of cutoff $\lambda c = 0.8$ mm, so the center of the filter is at the center of a profile of length 8 mm (spacing = 1 μm) ($k = 4{,}000$). Generate the filter over the entire length of the profile (assume l = 8 mm, Δx = 1 μm). What is the sum of the weights of this filter (before normalizing to unit sum)? What is the sum of the weights of the filter if the center of the filter window were at the first point of the profile? Compute the ratio of the two sums and verify that the ratio of the first points of the filtered profiles in Example 9.2 [ratio of $w_2(1)$ and $w_1(1)$] is identical to that computed above. Explain why that is so.

9.2. Suggest an implementation of the zero-order Gaussian regression filter that will reduce the computational time compared to the method shown in Example 9.2. Use the fact that the filter behaves identically with the Gaussian filter, except at the edges.

9.3. Generate a surface z containing 128×128 points, that has a sinusoidal wavelength along x of amplitude 2 μm and wavelength 6.4 mm. Assume spacing is

0.1 mm. Filter this surface using the 3D Gaussian filter and 3D zero-order Gaussian regression filter. Plot the results and review the performance of each filter near the edges. Let cutoff = 6.4 mm. Compare the results with that obtained in Example 8.3.

9.4. Determine the partial differentials with respect to the unknowns in Eq. 9.15 and develop the matrix form in a manner similar to that in Eq. 9.9.

References

Brinkmann, S., Bodschwinna, H. and Lemke, H.W. 2000, 'Development of a robust Gaussian regression filter for three-dimensional surface analysis', *Proceedings of the X International Colloquium on Surfaces, Chemnitz University of Technology. Chemnitz*, pp. 122–132.

Brinkmann, S., Bodschwinna, H. and Lemke, H.W. 2001, 'Accessing roughness in three dimensions using Gaussian regression filtering', *International Journal of Machine Tools and Manufacture*, vol. 41, no. 13–14, pp. 2153–2161.

Seewig, J. 2005, 'Linear and robust Gaussian regression filters', *Journal of Physics: Conference Series*, vol. 13, pp. 254–257.

Chapter 10
Spline Filter

10.1 Introduction

We focus attention here on a filter that promises to be the preferred option for a variety of applications because of its superior performance in comparison to the Gaussian filter. The spline filter proposed by Krystek (1996a, 1996b, 2000) and incorporated into ISO Standards (International Organization for Standardization 2006) is another solution to overcome the two problems associated with the Gaussian filter discussed previously, namely, edge distortion and poor performance in the presence of large form. There are two types of spline filters: non-periodic and periodic. Non-periodic spline filters are used for filtering open profiles (flat surfaces) and periodic spline filters for closed profiles (roundness).

Equations in this chapter are reproduced from Krystek (2000), with permission. The weighting function of a low-pass spline filter cannot be given by a simple closed formula. Therefore, filter equations are used instead of weighting functions to describe spline filters. The filter equation for the non-periodic spline filter is given by

$$(1+\alpha^4 Q)w = z\,, \tag{10.1}$$

where w is the filtered waviness profile containing n points, z is the raw profile also containing n points, and Q is a $n \times n$ matrix defined as

$$Q = \begin{bmatrix} 1 & -2 & 1 & & & & & \\ -2 & 5 & -4 & 1 & & & & \\ 1 & -4 & 6 & -4 & 1 & & & \\ & \cdot & \cdot & \cdot & \cdot & \cdot & & \\ & & \cdot & \cdot & \cdot & \cdot & \cdot & \\ & & & \cdot & \cdot & \cdot & \cdot & \cdot \\ & & & 1 & -4 & 6 & -4 & 1 \\ & & & & 1 & -4 & 5 & -2 \\ & & & & & 1 & -2 & 1 \end{bmatrix} \tag{10.2}$$

for non-periodic splines and

$$Q = \begin{bmatrix} 6 & -4 & 1 & & & & & & 1 & -4 \\ -4 & 6 & -4 & 1 & & & & & & 1 \\ 1 & -4 & 6 & -4 & 1 & & & & & \\ & & \cdot & \cdot & \cdot & \cdot & \cdot & & & \\ & & & \cdot & \cdot & \cdot & \cdot & \cdot & & \\ & & & & \cdot & \cdot & \cdot & \cdot & \cdot & \\ & & & & & 1 & -4 & 6 & -4 & 1 \\ 1 & & & & & & 1 & -4 & 6 & -4 \\ -4 & 1 & & & & & & 1 & -4 & 6 \end{bmatrix} \tag{10.3}$$

for periodic splines, and α is given by

$$\alpha = \left(2\sin\left(\frac{\pi\Delta x}{\lambda c}\right)\right)^{-1} . \tag{10.4}$$

10.2 Amplitude Transmission Characteristics

Because the low-pass spline filter does not have a weighting function, the transmission characteristics cannot be obtained from the DFT. In order to obtain the transmission characteristics, consider an input sinusoidal profile given by $z = A_0 \sin(2\pi x/\lambda)$. Because the spline filter has zero phase characteristics, let the output sinusoidal profile be given by $w = A_1 \sin(2\pi x/\lambda)$. Then, the amplitude transmission is A_1/A_0.

The jth point of array z is given by $z = A_0 \sin(2\pi(j-1).\Delta x/\lambda)$, and is related to the $j-2$, $j-1$, j, $j+1$, and $j+2$ points of array w by the spline filter equation. Therefore,

$$\alpha^4 w(j-2) - 4\alpha^4 w(j-1) + (1+6\alpha^4)w(j) - 4\alpha^4 w(j+1) + \alpha^4 w(j+2) = z(j) . \tag{10.5}$$

Substituting for w and z in Eq. 10.5, we get

$$\begin{aligned} &\alpha^4 A_1 \sin(2\pi(j-3).\Delta x/\lambda) - 4\alpha^4 A_1 \sin(2\pi(j-2).\Delta x/\lambda) \\ &+(1+6\alpha^4)A_1 \sin(2\pi(j-1).\Delta x/\lambda) - 4\alpha^4 A_1 \sin(2\pi(j).\Delta x/\lambda) \\ &+\alpha^4 A_1 \sin(2\pi(j+1).\Delta x/\lambda) = A_0 \sin(2\pi(j-1).\Delta x/\lambda) . \end{aligned} \tag{10.6}$$

Simplifying Eq. 10.6, we obtain the low-pass transmission characteristics as

$$\frac{A_1}{A_0} = \left(1 + 16\alpha^4 \sin^4\left(\frac{\pi\Delta x}{\lambda}\right)\right)^{-1} . \tag{10.7}$$

10.3 Implementation of the Non-periodic Spline Filter

The spline equation given in Eq. 10.1 can be written in matrix form $Aw = z$, where

$$A = \begin{bmatrix} 1+\alpha^4 & -2\alpha^4 & \alpha^4 & & & & & \\ -2\alpha^4 & 1+5\alpha^4 & -4\alpha^4 & \alpha^4 & & & & \\ \alpha^4 & -4\alpha^4 & 1+6\alpha^4 & -4\alpha^4 & \alpha^4 & & & \\ & \cdot & \cdot & \cdot & \cdot & \cdot & & \\ & & \cdot & \cdot & \cdot & \cdot & \cdot & \\ & & & \cdot & \cdot & \cdot & \cdot & \cdot \\ & & & \alpha^4 & -4\alpha^4 & 1+6\alpha^4 & -4\alpha^4 & \alpha^4 \\ & & & & \alpha^4 & -4\alpha^4 & 1+5\alpha^4 & -2\alpha^4 \\ & & & & & \alpha^4 & -2\alpha^4 & 1+\alpha^4 \end{bmatrix}. \tag{10.8}$$

There are n equations and n unknowns. Instead of solving them as simultaneous equations, matrix A can be first transformed into an upper triangular matrix using row manipulations. For example, when the second row is replaced as the difference between the second row and $\frac{-2\alpha^4}{1+\alpha^4}$ times the first row, the first column in the second row drops to zero. When the third row is replaced as the difference between the third row and $\frac{\alpha^4}{1+\alpha^4}$ times the first row, the first column in the third row also drops to zero. Such row manipulations eventually lead to an upper triangular matrix A, which in turn leads to an elegant solution for the elements of the waviness profile w.

Another point about the implementation is worth mentioning. Because typical surface profiles are long, matrix A becomes gigantic and care must be taken to address potential memory problems. Instead of storing the entire matrix A, the sparse nature of A is exploited and only non-zero elements of A are actually stored. We explain storage in detail for the periodic spline filter in Sect. 10.4.

Example 10.1. Plot the transmission characteristics of the spline filter. Assume cutoff $\lambda c = 0.8$ mm, spacing $\Delta x = 1$ µm. Also plot the transmission characteristics of the Gaussian filter for comparison.

The following MATLAB code generates the transmission characteristics of the spline and Gaussian filter. Notice that the spline has a steeper transmission in comparison to the Gaussian (see Fig. 10.1).

```
% SPLINE
lambdac = 0.8;                        % lambdac in mm
dx = 0.001;                           % spacing in mm
alpha = 1/(2*sin(pi*dx/lambdac));
lambda = (0.002:0.001:8)';            % generate wavelength axis
                                      % for transmission plot
T = 1./(1+16*alpha^4*sin(pi*dx./lambda).^4); % sample the
                                      % spline transmission equation

semilogx(lambda,T*100);
xlabel('Wavelength (mm)');
```

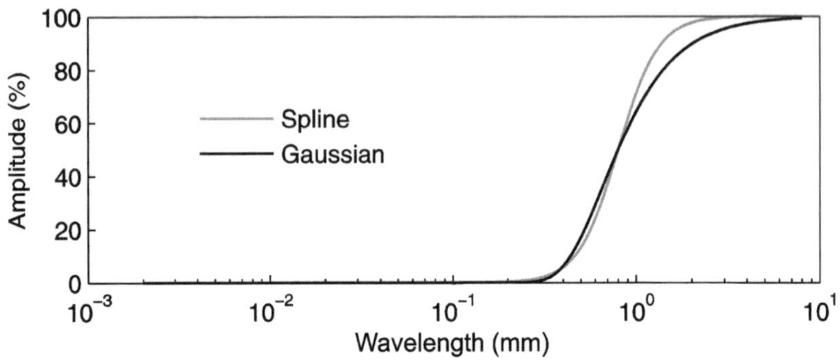

Fig. 10.1 Transmission characteristics of the spline and Gaussian filters

```
ylabel('Amplitude (%)');
hold on
% GAUSSIAN
alpha = 0.4697;
lambda = (0.002:0.001:8)'; % generate wavelength axis
                           % for transmission plot
G = exp(-pi*(alpha*lambdac./lambda).^2); % sample the Gaussian
                                         % transmission equation
semilogx(lambda,G*100,'r');
xlabel('Wavelength (mm)');
ylabel('Amplitude (%)');
```

Steeper transmission characteristics are generally preferred, so that the partition between any two wavelength bands is more precise. A step transmission (illustrated in Chap. 3) will be an ideal filter for surface metrology, but is rarely used because it has no time-domain representation. That is, a step transmission characteristic curve has "rings" or lobes in the time domain and is therefore not suitable.

Example 10.2. Write a MATLAB function that implements the spline filter. The input arguments are the raw profile, spacing, and the cutoff.

Profile z is a column vector. Profile z, spacing dx, and cutoff *lambdac* are in the same units. The general idea here is to store elements of matrix A in different arrays (see implementation details in Sect. 10.4) for optimal storage purposes, and then to create an upper triangular matrix from which elements of w can be determined.

```
function [x,w] = npspline(z, dx, lambdac)
n = size(z,1);
x = (0:n-1)'*dx;
p = (0.5/sin(pi*dx/lambdac))^4;
f = 1 + 6*p;
g = -4*p;
h = p;
for k = 1:n,
    a(k) = f;
```

```
    b(k) = g;
    c(k) = h;
end
a(n) = a(n) - 5*p;
a(1) = a(n);
b(n-1) = b(n-1) + 2*p;
b(1) = b(n-1);
a(n-1) = a(n-1) - p;
a(2) = a(n-1);
for i = 1:(n-2),
    h = b(i)/a(i);
    a(i+1) = a(i+1) - h*b(i);
    b(i+1) = b(i+1) - h*c(i);
    z(i+1,1) = z(i+1,1) - h*z(i,1);
    h = c(i)/a(i);
    a(i+2) = a(i+2) - h*c(i);
    z(i+2,1) = z(i+2,1) - h*z(i,1);
end
h = b(n-1)/a(n-1);
a(n) = a(n) - h*b(n-1);
w(n,1) = (z(n,1) - h*z(n-1,1))/a(n);
w(n-1,1) = (z(n-1,1) - b(n-1) * w(n,1))/a(n-1);
for j=(n-2):-1:1,
    w(j,1) = (z(j,1) - b(j)*w(j+1,1) - c(j)*w(j+2,1))/a(j);
end
```

Example 10.3. Filter profile z in Example 9.2 using the spline and the Gaussian filter. Assume $\lambda c = 0.8$ mm and sampling interval $\Delta x = 1$ µm.

Continuing from Example 9.2, the spline waviness profile is generated and plotted along with profiles z and Gaussian waviness profile w_1 as shown in Fig. 10.2. The spline filter shows no edge distortion. Also, zooming into the central portion of the profile, it can be seen that the spline waviness profile rides above the Gaussian waviness profile, indicating that it is less influenced by the overall form and not weighted down.

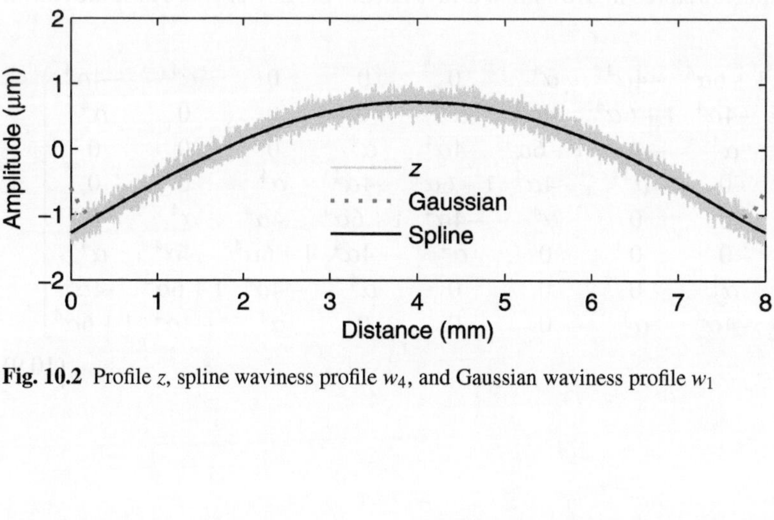

Fig. 10.2 Profile z, spline waviness profile w_4, and Gaussian waviness profile w_1

```
lambdac = 0.8;                          % cutoff in mm
dx = 0.001;                             % spacing in mm
[x,w4] = npspline(z, dx, lambdac); % generate spline
                                        % waviness profile

plot(x,z,'g',x,w4,'r',x,w1,'b')
xlabel('Distance (mm)');
ylabel('Amplitude (\mum)');
```

10.4 Implementation of the Periodic Spline Filter

While the periodic spline filter is very similar in its definition to the non-periodic spline filter, its implementation requires one additional step. In the case of the non-periodic spline filter, we were able to convert matrix A into an upper triangular matrix. In this case, A is positive definite and therefore we use Cholesky decomposition to express A as $R^T R$ where R is an upper triangular matrix. Then, the spline equation can be written as $R^T Rw = z$. Let $Rw = P$. Then, we first solve $R^T P = z$ for P by forward substitution. Finally, we solve $Rw = P$ by backward substitution.

As with the non-periodic spline filter, because A is sparse, we store the elements of A using several row matrices. We explain this in more detail here. An example for a data set containing eight points is described here. It can be seen that the matrix is symmetric and therefore, we are only required to store elements on or above the main (top-left to bottom-right) diagonal. Second, we note that the elements are concentrated in a small band near the diagonal and there are three elements in the far right. Therefore we define five arrays: a, b, c, d, and e. Array a contains elements of the main diagonal, b contains the first off-diagonal elements, and c contains the second off-diagonal elements. Arrays d and e store the last two columns. Array a is thus $n-2$ elements long, b is $n-3$ long, c is $n-4$ long, d is $n-1$ long, and e is n elements long. Also, while the last two columns of A are mostly zero, Cholesky decomposition will fill most of the last two columns. The results of Cholesky decomposition are stored in five similar arrays La, Lb, Lc, Ld, and Le. These are shown below:

$$A = \begin{bmatrix} 1+6\alpha^4 & -4\alpha^4 & \alpha^4 & 0 & 0 & 0 & \alpha^4 & -4\alpha^4 \\ -4\alpha^4 & 1+6\alpha^4 & -4\alpha^4 & \alpha^4 & 0 & 0 & 0 & \alpha^4 \\ \alpha^4 & -4\alpha^4 & 1+6\alpha^4 & -4\alpha^4 & \alpha^4 & 0 & 0 & 0 \\ 0 & \alpha^4 & -4\alpha^4 & 1+6\alpha^4 & -4\alpha^4 & \alpha^4 & 0 & 0 \\ 0 & 0 & \alpha^4 & -4\alpha^4 & 1+6\alpha^4 & -4\alpha^4 & \alpha^4 & 0 \\ 0 & 0 & 0 & \alpha^4 & -4\alpha^4 & 1+6\alpha^4 & -4\alpha^4 & \alpha^4 \\ \alpha^4 & 0 & 0 & 0 & \alpha^4 & -4\alpha^4 & 1+6\alpha^4 & -4\alpha^4 \\ -4\alpha^4 & \alpha^4 & 0 & 0 & 0 & \alpha^4 & -4\alpha^4 & 1+6\alpha^4 \end{bmatrix} \tag{10.9}$$

$$A = \begin{bmatrix} a(1) & b(1) & c(1) & 0 & 0 & 0 & d(1) & e(1) \\ 0 & a(2) & b(2) & c(2) & 0 & 0 & d(2) & e(2) \\ 0 & 0 & a(3) & b(3) & c(3) & 0 & d(3) & e(3) \\ 0 & 0 & 0 & a(4) & b(4) & c(4) & d(4) & e(4) \\ 0 & 0 & 0 & 0 & a(5) & b(5) & d(5) & e(5) \\ 0 & 0 & 0 & 0 & 0 & a(6) & d(6) & e(6) \\ 0 & 0 & 0 & 0 & 0 & 0 & d(7) & e(7) \\ 0 & 0 & 0 & 0 & 0 & 0 & 0 & e(8) \end{bmatrix} \quad (10.10)$$

$$R = \begin{bmatrix} La(1) & Lb(1) & Lc(1) & 0 & 0 & 0 & Ld(1) & Le(1) \\ 0 & La(2) & Lb(2) & Lc(2) & 0 & 0 & Ld(2) & Le(2) \\ 0 & 0 & La(3) & Lb(3) & Lc(3) & 0 & Ld(3) & Le(3) \\ 0 & 0 & 0 & La(4) & Lb(4) & Lc(4) & Ld(4) & Le(4) \\ 0 & 0 & 0 & 0 & La(5) & Lb(5) & Ld(5) & Le(5) \\ 0 & 0 & 0 & 0 & 0 & La(6) & Ld(6) & Le(6) \\ 0 & 0 & 0 & 0 & 0 & 0 & Ld(7) & Le(7) \\ 0 & 0 & 0 & 0 & 0 & 0 & 0 & Le(8) \end{bmatrix} . \quad (10.11)$$

After allocating appropriate storage, the next step in the implementation is to compute the terms of the Cholesky decomposition matrix R. Because $R^T R = A$, we can find the terms of R iteratively as follows. The first elements of each array are given by

$$\begin{aligned} La(1) &= \sqrt{a(1)} \\ Lb(1) &= b(1)/La(1) \\ Lc(1) &= c(1)/La(1) \\ Ld(1) &= d(1)/La(1) \\ Le(1) &= e(1)/La(1) \end{aligned} . \quad (10.12)$$

The second elements are then given by

$$\begin{aligned} La(2) &= \sqrt{a(2) - Lb(1)^2} \\ Lb(2) &= (b(2) - Lb(1)Lc(1))/La(2) \\ Lc(2) &= c(2)/La(2) \\ Ld(2) &= (d(2) - Lb(1)Ld(1))/La(2) \\ Le(2) &= (e(2) - Lb(1)Le(1))/La(2) \end{aligned} . \quad (10.13)$$

Subsequent elements are then found iteratively. After computing the Cholesky decomposition, we solve the system $R^T P = z$. Because R^T is a lower triangular matrix, the system can be easily solved. The first few elements of P are given by

$$\begin{aligned} P(1) &= z(1)/La(1) \\ P(2) &= (z(2) - Lb(1)z(1))/La(2) \\ P(3) &= (z(3) - Lc(1)z(1) - Lb(2)z(2))/La(3) \end{aligned} . \quad (10.14)$$

These are then used in an iterative fashion to determine the remaining elements of P. Finally, we solve the system $Rw = P$ in a similar manner, but because R is upper triangular, we first solve for $w(n)$ and then proceed backwards. The solution strategy described above and the algorithm in the following example are reproduced (with modifications) from Krystek (1996b) with permission.

Example 10.4. Write a MATLAB program that implements the periodic spline filter.

The program is shown below. The inputs are the profile points and the cutoff in undulations per revolution (UPR).

```
% z is a column vector of data, omegac is the cutoff in UPR
function [x,r,w] = pspline(z,omegac)
n = size(z,1);
x = (0:1:n-1)'*2*pi/n;
alpha = 1/2/sin(pi*omegac/n);
alpha4 = alpha^4;
% create vectors a, b, c, d and e
for i=1:n
    a(i) = 1+6*alpha4;
    b(i) = -4*alpha4;
    c(i) = alpha4;
    d(i) = 0; e(i) = 0;
end
d(1) = alpha4;
d(n-3) = alpha4;
d(n-2) = -4*alpha4;
d(n-1) = 1+6*alpha4;
d(n) = -4*alpha4;
e(1) = -4*alpha4;
e(2) = alpha4;
e(n-2) = alpha4;
e(n-1) = -4*alpha4;
e(n) = 1+6*alpha4;
% compute vectors La, Lb, Lc, Ld and Le
La(1) = sqrt(a(1));
Lb(1) = b(1)/La(1);
Lc(1) = c(1)/La(1);
Ld(1) = d(1)/La(1);
Le(1) = e(1)/La(1);
La(2) = sqrt(a(2) - Lb(1)*Lb(1));
Lb(2) = (b(2) - Lb(1)*Lc(1))/La(2);
Lc(2) = c(2)/La(2);
Ld(2) = (d(2) - Lb(1)*Ld(1))/La(2);
Le(2) = (e(2) - Lb(1)*Le(1))/La(2);
for i=3:n-4
    La(i) = sqrt(a(i) - Lb(i-1)*Lb(i-1) - Lc(i-2)*Lc(i-2));
    Lb(i) = (b(i) - Lb(i-1)*Lc(i-1))/La(i);
    Lc(i) = c(i) /La(i);
    Ld(i) = (d(i) - Lc(i-2)*Ld(i-2)-Lb(i-1)*Ld(i-1))/La(i);
    Le(i) = (e(i) - Lc(i-2)*Le(i-2)-Lb(i-1)*Le(i-1))/La(i);
end
i=n-3;
La(i) = sqrt(a(i) - Lb(i-1)*Lb(i-1) - Lc(i-2)*Lc(i-2));
Lb(i) = (b(i) - Lb(i-1)*Lc(i-1))/La(i);
```

```
Ld(i) = (d(i) - Lc(i-2)*Ld(i-2)-Lb(i-1)*Ld(i-1))/La(i);
Le(i) = (e(i) - Lc(i-2)*Le(i-2)-Lb(i-1)*Le(i-1))/La(i);
i=n-2;
La(i) = sqrt(a(i) - Lb(i-1)*Lb(i-1) - Lc(i-2)*Lc(i-2));
Ld(i) = (d(i) - Lc(i-2)*Ld(i-2)-Lb(i-1)*Ld(i-1))/La(i);
Le(i) = (e(i) - Lc(i-2)*Le(i-2)-Lb(i-1)*Le(i-1))/La(i);
i=n-1;
temp1 = 0; temp2 = 0;
for j=1:n-2
    temp1 = temp1 + Ld(j)*Ld(j);
    temp2 = temp2 + Ld(j)*Le(j);
end
Ld(i) = sqrt(d(i) - temp1);
Le(i) = (e(i) - temp2)/Ld(i);
i=n;
temp1 = 0;
for j=1:n-1
    temp1 = temp1 + Le(j)*Le(j);
end
Le(i) = sqrt(e(i) - temp1);
% Now Solve the System of Equations
P(1) = z(1)/La(1);
P(2) = (z(2) - Lb(1)*P(1))/La(2);
P(3) = (z(3) - Lc(1)*P(1)-Lb(2)*P(2))/La(3);
for i=4:n-2
    P(i) = (z(i) - Lc(i-2)*P(i-2)-Lb(i-1)*P(i-1))/La(i);
end
temp = 0;
for i=1:n-2
    temp = temp + Ld(i)*P(i);
end
P(n-1) = (z(n-1) - temp)/Ld(n-1);
temp = 0;
for i=1:n-1
    temp = temp + Le(i)*P(i);
end
P(n) = (z(n) - temp)/Le(n);
w(n,1) = P(n)/Le(n);
w(n-1,1) = (P(n-1) - Le(n-1)*w(n,1))/Ld(n-1);
w(n-2,1) = (P(n-2) - Ld(n-2)*w(n-1,1) - Le(n-2)*w(n,1))/La(n-2);
w(n-3,1) = (P(n-3) - Lb(n-3)*w(n-2,1) - Ld(n-3)*w(n-1,1) -
Le(n-3)*w(n,1))/La(n-3);
w(n-4,1) = (P(n-4) - Lb(n-4)*w(n-3,1) - Lc(n-4)*w(n-2,1)
- Ld(n-4)*w(n-1,1) - Le(n-4)*w(n,1))/La(n-4);
for i=n-5:-1:1
    w(i,1) = (P(i) - Lb(i)*w(i+1,1) - Lc(i)*w(i+2,1)
    - Ld(i)*w(n-1,1) - Le(i)*w(n,1))/La(i);
end
r = z - w;
```

10.5 Summary

We have discussed the periodic and non-periodic spline filter in this chapter. The spline filter is a relatively new filtering technique. It has several advantages over the Gaussian filter: it has steeper transmission, overcomes edge effects, and performs better in the presence of large form. Its primary disadvantage is that there is no expression for the weighting function and is therefore cumbersome to implement. The spline filter has been incorporated into the ISO standards (ISO/TS 16610-22:2006) with minor modifications in the spline equation.

Exercises

10.1. Simplify Eq. 10.5 and show that the transmission characteristic is given by Eq. 10.7.

10.2. Filter profile z in Example 9.2 using the following filters: (a) a spline filter, (b) a Gaussian filter, (c) a zero-order Gaussian regression filter, and (d) a second-order Gaussian regression filter. Use cutoff of 0.8 mm. Plot the waviness profiles generated using the four filters on a single chart. Qualitatively assess the performance of the filters at the edges of the profile. Quantitatively assess the speed of the filtering operation for each filter.

10.3. Filter a sinusoidal profile z of amplitude 1 μm and wavelength 0.8 mm (8 mm long with spacing 1 μm) using a Gaussian and a spline filter of cutoff 0.8 mm. Why does the Gaussian waviness profile appear to have no edge effects while the spline does? Repeat the process using a cutoff of 0.08 mm. Explain the observations.

10.4. Repeat the above simulations, but with a profile that is 7.6 mm long, all other parameters being the same. Explain the results.

10.5. Filter roundness profile in Example 7.3 using the periodic spline filter with 50 UPR cutoff. Plot the raw profile and the waviness profile on a polar plot. Filter the same profile using the non-periodic spline, assuming the part radius is 0.573 mm. Is there any difference between the two waviness profiles?

References

International Organization for Standardization 2006, *ISO/TS 16610-22:2006, Geometrical Product Specifications (GPS) – Filtration – Part 22: Linear Profile Filters: Spline Filters*, ISO. Geneva, Switzerland.

Krystek, M. 1996a, 'Form filtering by splines', *Measurement*, vol. 18, pp. 9–15.

Krystek, M. 1996b, 'Discrete L-spline filtering in roundness measurements', *Measurement*, vol. 18, no. 2, pp. 129–138.

Krystek, M. 2000, 'Discrete linear profile filters', *Proceedings of the X International Colloquium on Surfaces, Chemnitz University of Technology, Chemnitz*, pp.145–152.

Chapter 11
Robust Filters

11.1 *Rk* Filter

The Gaussian filter waviness profile is not robust in the presence of outliers, such as a scratch in the surface, valleys generated during machining, dirt, burrs, *etc*. That is, the waviness profile follows these outliers to some extent, resulting in artificial features being introduced in the roughness profile.

There are several solutions to overcome this problem. The simplest procedure is referred to as *Rk* filtering, and was originally developed to filter stratified surfaces such as those produced in plateau-honing operation. Plateau-honed surfaces have a core texture superimposed on deep valleys. The filtering operation proceeds as follows. The primary profile is filtered using a low-pass filter. All points in the primary profile that lie below the waviness profile are replaced by the waviness profile itself to generate a modified profile. The modified profile is then filtered again using the previous low-pass filter to obtain the waviness profile. This waviness profile is subtracted from the original profile to obtain the roughness profile.

The filter used can be the Gaussian, the spline or even the triangular. The ISO Standard ISO 13565-1:1996 (E) (International Organization for Standardization 1996) recommends the use of the Gaussian filter, while the German Deutsches Institut für Normung (DIN) Standard, DIN 4776:1990 (Deutsches Institut für Normung 1990), recommends the triangular filter. However, the more recently proposed spline filter could prove to be a better choice because it has no edge effects; the entire profile can therefore be used for parameter assessment, as opposed to the Gaussian or the triangular filter, where a considerable portion of the profile has to be discarded.

Example 11.1. Generate a sinusoidal profile z (amplitude 1 μm, wavelength 16 mm), in the interval $0 \leqslant x < 8$ mm, sampled at a spacing of 1 μm. Superimpose normally distributed random noise of zero mean and 0.1 μm standard deviation. Also superimpose valleys of depth 3 μm and width 100 μm at two locations, $x = 2$ mm and $x = 3$ mm, to simulate outliers. Filter profile z using the Gaussian filter of 0.8 mm cutoff. Plot both the waviness and roughness profiles. Is the Gaussian waviness profile robust against outliers?

The following MATLAB code generates profile z and plots the waviness and roughness profiles. Notice that the waviness profile is weighted down near the valleys, as shown in Fig. 11.1. This generates artificial features in the roughness profile (the raised portions immediately prior to and after a valley). The Gaussian waviness profile is not robust against outliers.

```
n = 8000;                     % number of profile points
dx = 0.001;                   % spacing in mm
x = (0:1:n-1)'*dx;            % generate x axis
z = detrend(1*sin(2*pi.*x/16)+normrnd(0,0.1*ones(n,1)));
                              % generate profile z
z(2001:2100,1) = -3;          % superimpose valley of depth 3 um
                              % from 2 mm to 2.1 mm
z(3001:3100,1) = -3;          % superimpose valley of depth 3 um
                              % from 3 mm to 3.1 mm
% GAUSSIAN FILTER
lambdac = 0.8;                % cutoff in mm
xg = (-lambdac:dx:lambdac)'; % generate x axis for filter
alpha = 0.4697;
S = (1/(alpha*lambdac)).*exp(-pi*(xg/(alpha*lambdac)).^2);
% Gaussian filter array
S = S/sum(S);                 % normalize to unit sum
w1 = conv(S,z);               % perform convolution
w1 = w1(801:8800,1);          % extract central n points
r1 = z - w1;                  % determine roughness profile
plot(x,z,'g',x,w1,'r');
xlabel('Distance (mm)');
ylabel('Amplitude (\mum)');
plot(x,r1);
xlabel('Distance (mm)');
ylabel('Amplitude (\mum)');
```

Example 11.2. Filter profile z in Example 11.1 using the two-step *Rk* filtering approach. Use the Gaussian weighting function at 0.8 mm cutoff.

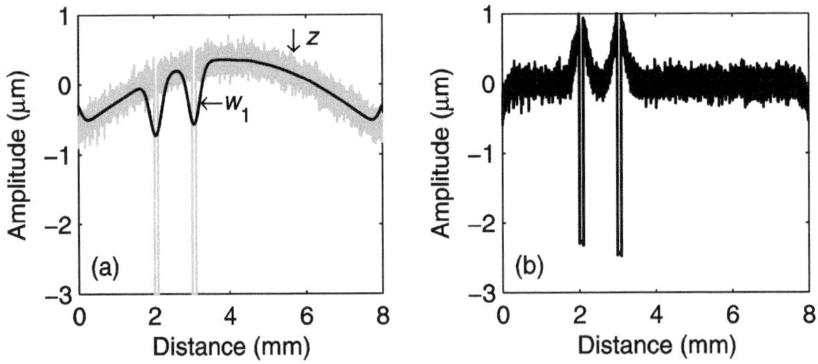

Fig. 11.1 **a** Raw profile z and Gaussian waviness profile w_1. **b** Roughness profile

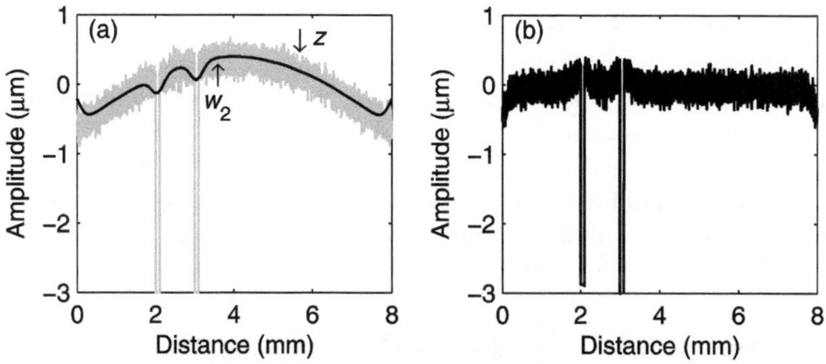

Fig. 11.2 **a** Raw profile z and waviness profile w_2 generated using the two-step *Rk* filtering with the Gaussian filter. **b** Roughness profile

The two-step procedure is implemented below. The waviness and roughness profiles are shown in Fig. 11.2. While the waviness profile shows considerable improvement near the valleys, the distortion is not completely removed. Conceivably, multiple passes of the Gaussian filter will be more effective. In that case, there is a need for termination criteria to stop the iterations. Such multiple-pass filters are discussed in Sect. 11.2. Continuing from Example 11.1, we have the following:

```
% GAUSSIAN RK FILTERING METHOD
% FIRST PASS
tmp = conv(z,S);               % convolve profile z and filter S
tmp = tmp(801:8800,1);         % extract central n points
zc = z;                        % create a copy of profile z
% Replace all points in raw profile z below waviness profile
% by the waviness profile itself
for i=1:n
    if (zc(i,1) < tmp(i,1))   % if raw profile point i is below
                               % waviness profile point i
        zc(i,1) = tmp(i,1);   % then, replace profile point i by
                               % waviness point i
    end
end
% SECOND PASS CONVOLUTION
tmp = conv(zc,S);              % convolve profile zc and filter S
w2 = tmp(801:8800,1);          % extract central n points
r2 = z - w2;                   % determine roughness profile
plot(x,z,'g',x,w2,'r');
xlabel('Distance (mm)');
ylabel('Amplitude (\mum)');
plot(x,r2);
xlabel('Distance (mm)');
ylabel('Amplitude (\mum)');
```

11.2 Robust Gaussian Regression Filter

The robust regression filter is an improvement over the two-step *Rk* filtering procedure in that multiple passes of the Gaussian filter are applied to the profile. The termination is based on a median statistic. In order to avoid edge effects, a regression filter is applied instead of the traditional Gaussian filter. The robust Gaussian regression filter (currently a draft international Standard) is given by the following arrangement (Brinkmann *et al.*, 2000):

$$\text{Minimize}\, E(k) = \sum_{p=1}^{n} (z(p) - w(k))^2 \delta(i,p) S(k,p) \Delta x \,, \qquad (11.1)$$

where $\delta(i,p)$ is an additional vertical weight and i is the iteration number. The other terms are defined in Chap. 9. In the first iteration of the filter, $\delta(i,p)$ is set to unity. In subsequent iterations, $\delta(i,p)$ is set according to

$$\delta(i+1,p) = \left(1 - \left(\frac{r(i,p)}{C_B}\right)^2\right)^2 \quad \text{for} \left|\frac{r(i,p)}{C_B}\right| < 1 \,, \qquad (11.2)$$

and $\delta(i+1,p) = 0$ otherwise.

C_B is computed as $4.4 \times \text{Median}(|r(i,p)|)$, where $r(i,p) = z(p) - w(p)$.

Equations 11.1 and 11.2 are reproduced from Brinkmann *et al.* (2000) with permission.

Example 11.3. Apply the robust zero-order Gaussian regression filter on profile z in Example 11.1. Compare the waviness profiles with that obtained using the Gaussian and *Rk* filters in Examples 11.1 and 11.2.

The following MATLAB code generates the robust zero-order Gaussian regression waviness and roughness profiles (see Fig. 11.3).

```
const = sqrt(log(2)/2/pi/pi);
iterationNo = 0;          % set current iteration number to zero
delta = ones(n,1);        % array related to robust
                          % Gaussian regression filter

CB = 1;
while CB > 0.304          % CB threshold by trial and error
    for k=1:n
        p=(1:1:n)';       % generate x array for filter
        S = (1/sqrt(2*pi)/const/lambdac).*exp(-0.5*((k-p)
        *dx/const/lambdac).^2);
                          % generate weighting function
        S_modified = S/sum(S); % modified weighting function
        w3(k,1) = sum(z.*S_modified.*delta)/sum(S_modified.*delta);
                          % compute mean line
    end
    r3 = z-w3;            % roughness profile
    CB = 4.4*median(abs(r3)); % compute CB using median statistic
    for j=1:n
```

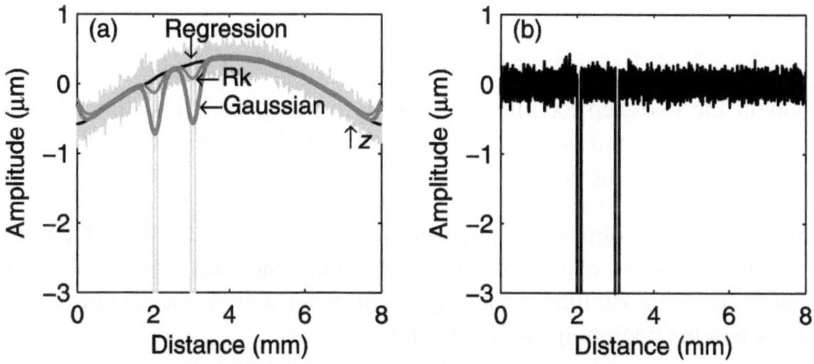

Fig. 11.3 **a** Raw profile *z* and waviness profiles generated using the robust zero-order Gaussian regression filter, two-step *Rk* filtering with the Gaussian filter, and traditional Gaussian filter. **b** Roughness profile obtained using the robust zero-order Gaussian regression filter

```
        if (abs(r3(j,1)/CB) < 1)     % check condition
                                     % given in Eq. 11.2
             delta(j,1) = (1-(r3(j,1)/CB).^2).^2;
        else delta(j,1)= 0;
        end
    end
    iterationNo = iterationNo + 1; % increment iteration number
end
plot(x,z,'g',x,w3,'r',x,w2,'b',x,w1,'k');
xlabel('Distance (mm)');
ylabel('Amplitude (\mum)');
plot(x,r3);
xlabel('Distance (mm)');
ylabel('Amplitude (\mum)');
```

11.3 Summary

Robust filters are designed not to be influenced by outliers in data sets. In an industrial measurement scenario, a measurement area of a surface can have pieces of dirt despite careful cleaning. Because it may be time consuming to measure the profile again, analysis techniques that are robust against outliers are preferred. The two-step *Rk* procedure is quick and easy to implement, but not exceptionally robust. The robust zero-order Gaussian regression filter is computationally more intense, but handles outliers quite efficiently. Robust filters based on splines are also possible (Goto *et al.*, 2005), but not discussed here.

Exercises

11.1. Implement the two-step *Rk* filtering procedure using the spline filter. Filter profile z in Example 11.1 and plot the waviness and roughness profiles. Notice the superior performance near the edges.

11.2. The waviness profile obtained using a one-step Gaussian filter is distorted near the edges (a length equal to one cutoff in the beginning and end). If the Gaussian filter were used as in the *Rk* filtering arrangement, what will be the length of the distortion region in the beginning and in the end?

11.3. What is the effect of cutoff on robust filtering? In Example 11.1, will increasing the cutoff or decreasing the cutoff make the waviness profile more robust against outliers. Plot the waviness profiles for three different cutoffs: 0.25 mm, 0.8 mm and 2.5 mm.

References

Brinkmann, S., Bodschwinna, H. and Lemke, H.W. 2000, 'Development of a robust Gaussian regression filter for three-dimensional surface analysis', *Proceedings of the X International Colloquium on Surfaces, Chemnitz University of Technology. Chemnitz*, pp. 122–132.

Deutsches Institut für Normung 1990, *DIN 4776:1990, Measurement of Surface Roughness – Parameters* $R_k, R_{pk}, R_{vk}, M_{r1}, M_{r2}$ *for Describing the Material Portion in the Roughness Profile*, DIN.

Goto, T., Miyakura, J., Umeda, K., Kadowaki, S. and Yanagi, K. 2005, 'A robust spline filter on the basis of L2-norm', *Precision Engineering*, vol. 29, no. 2, pp. 157–161.

International Organization for Standardization 1996, *ISO 13565-1:1996 (E), Geometrical Product Specifications (GPS) – Surface Texture: Profile Method; Surfaces Having Stratified Functional Properties – Part 1: Filtering and General Measurement Condition*, ISO. Geneva, Switzerland.

Chapter 12
Envelope and Morphological Filters

12.1 Envelope Filters

An envelope filter is different from other filters discussed in previous chapters in that the mean line is not generated by an averaging process; rather the envelope generated by simulating the process of rolling a ball or sliding a plane over the surface is the mean line. The motivation for this approach is driven from a functional standpoint. Any number of applications can be identified where relative motion between two contacting surfaces is critical. In such cases, the envelope mean line is sometimes believed to yield a better representation of the contact phenomena.

As mentioned in Chap. 2, in the early days of filtering, the envelope system of filtering (the E system) competed against the frequency-domain (or averaging approach) filtering (the M system) method. Proposed by Von Weingraber (1956, 1957), the technique was researched extensively by Von Weingraber, Radhakrishnan, and Shunmugam (Radhakrishnan 1972; Shunmugam and Radhakrishnan 1974; Radhakrishnan and Von Weingarber 1969, 1971). The M system, came to be more generally accepted and for many decades has been the standard method of filtering surface profiles. The E system is once again finding new relevance, particularly with the introduction of morphological filtering in the international Standards ISO/TS 16610-40:2006 and ISO/TS 16610-41:2006 (International Organization for Standardization 2006a, 2006b). Morphological filtering is essentially a superset of the E system and offers more tools and capabilities. The renewed interest in metrology for such rolling-ball-type filters was championed by Srinivasan (1998) and Scott (2000), among others.

Consider the situation where a ball rolls over a surface as shown in Fig. 12.1. The ball is only allowed vertical translation, and contacts the proudest peaks as it traverses the surface. The locus of the center of the ball defines the envelope mean line. It is generated as shown below.

Let the ball be sampled along the x axis as shown in Fig. 12.1. The vertical position of the surface (bottom) of the ball from its center line is given by the array B. $B(5)$ is the central point and its value is equal to the radius. At any given profile

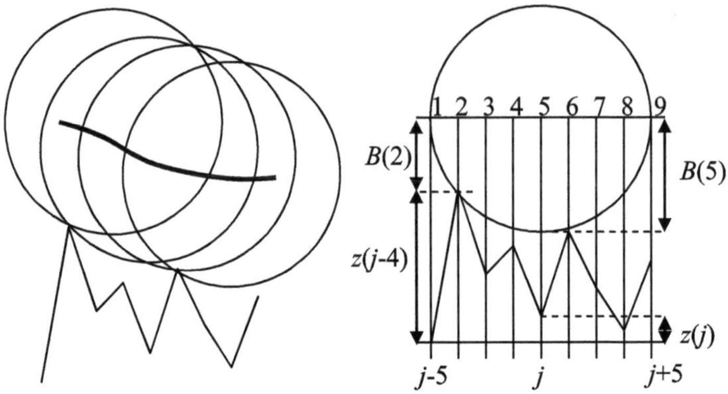

Fig. 12.1 Locus of the center of the ball as the ball traverses the surface

point, the vertical position of the center of ball is dependent not only on the profile point directly below the center, but also on the neighboring profile points that lie below the ball.

When the ball is centered on profile point j, the center of the ball is not necessarily determined as $z(j)+B(5)$. This is because a large neighboring profile peak can contact the ball on its curved surface and raise the center further. This is shown in Fig. 12.1, where a peak at $z(j-4)$ contacts the ball at three units to the left from the center and determines the location of the center of the ball. Therefore, to evaluate the locus of the center of the ball as it translates along the surface, the sum of profile heights and ball ordinates are evaluated over the region the ball overlaps the profile. The maximum of these values is the locus of the center of ball at that profile location j.

When the ball radius is large, the mean line overlooks the narrower peaks and valleys. On the other hand, when the radius is small, the mean line enters the narrow peaks and valleys. This is illustrated in Example 12.1.

12.2 Stylus Tip Convolution

When a stylus-based surface profile measuring instrument is used to obtain a surface profile, the raw data is limited at the high-frequency end by the finite size of the stylus. The raw data as recorded by the instrument is a complex function of the stylus geometry and surface asperity height, and spacing (Radhakrishnan 1970; Whitehouse 1974). To some extent, the measured profile can be approximated by the envelope mean line of the true profile, generated using a ball of size equal to the stylus diameter. Therefore, any surface features whose widths are narrower than the stylus size will not be captured by the instrument. Stylus tip convolution, as this effect is sometimes referred to, is not as much a problem in traditional surface metrology, as it is in nano-metrology, where extremely fine surface features are measured using sharp tips with instruments such as the atomic force microscope (AFM) (Villarrubia 1996).

Example 12.1. Generate a profile z simulated as the sum of the following three components defined in the interval $0 \leqslant x < 8$ mm, with a spacing $\Delta x = 1$ μm: (a) a sinusoid of amplitude 0.2 μm and wavelength 16 mm, (b) a sinusoid of amplitude 0.05 μm and wavelength 0.1 mm, and (c) normally distributed random noise of zero mean and 0.03 μm standard deviation. Generate the envelope mean lines using balls of radii 0.5 mm and 0.05 mm. Plot the results and explain. The following MATLAB code segment generates profile z.

```
dx = 0.001;                    % spacing in mm
n = 8000;                      % number of profile points
x = (0:n-1)'*dx;               % generate x axis
z = 0.2*sin(2*pi.*x/16) + 0.05*sin(2*pi.*x/0.1) +
normrnd(0,0.03*ones(n,1)); % generate profile z
```

The ball is then sampled at the same spacing as the profile. The length of the profile is extended on both sides (zero-padding) by a length equal to the radius of the profile so that when the ball is at the beginning or at the end of the profile, it still overlaps a profile element. Figure 12.2 shows the profile z along with the envelope mean lines generated using the two different balls.

```
% Generate ball of radius 0.5 mm
Radius = 0.5;                     % radius in mm
m = Radius/dx;                    % number of points in one radius
                                  % of the ball
xr = (-Radius:0.001:Radius)';     % generate x axis over an entire
                                  % diameter
                                  % (overlap area with profile)
B = sqrt(Radius^2-xr.^2);         % get vertical heights for each x
                                  % location
% Find mean line
z1 = [zeros(m,1); z ; zeros(m,1)]; % zeropad z so that it is
                                   % extended by one radius
```

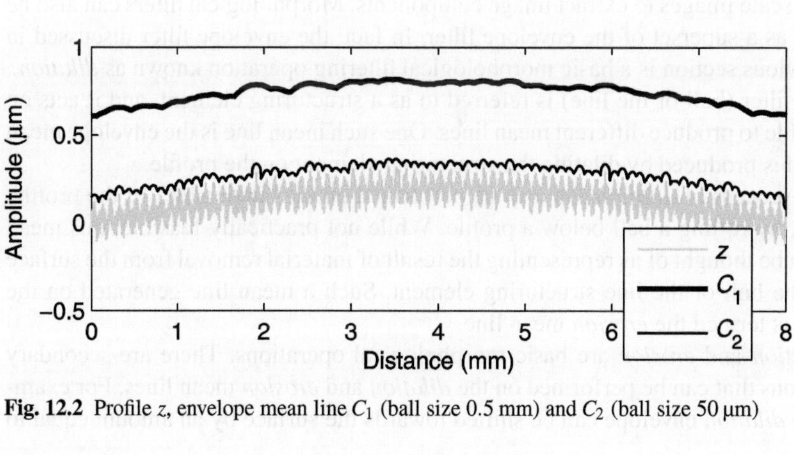

Fig. 12.2 Profile z, envelope mean line C_1 (ball size 0.5 mm) and C_2 (ball size 50 μm)

```
                                              % at both ends
for j=m+1:n+m
    C1(j-m,1) = max([z1(j-m:j+m,1)+B]); % compare sum of profile
                                         % and ball heights
end
% Generate ball of radius 0.05 mm
Radius = 0.05;                       % radius in mm
m = Radius/dx;                       % number of points in one radius
xr = (-Radius:0.001:Radius)';        % generate x axis over an entire
                                     % diameter
                                     % (overlap area with profile)
B = sqrt(Radius^2-xr.^2);            % get vertical heights for each x
                                     % location
% Find mean line
z1 = [zeros(m,1); z ; zeros(m,1)]; % zeropad z so that it is
                                    % extended by one radius
                                    % at both ends
for j=m+1:n+m
    C2(j-m,1) = max([z1(j-m:j+m,1)+B]); % compare sum of profile
                                         % and ball heights
end
% Plot
plot(x,z,'g',x,C1,'r',x,C2,'b');
xlabel('Distance (mm)');
ylabel('Amplitude (\mum)');
```

Notice that the mean line generated using the larger ball is smoother because the larger radius prevents it from entering the narrower valleys. Also, it is offset from the profile by a larger amount because of its larger radius.

12.3 Morphological Filters

Morphological filters are used extensively in digital image processing (Gonzales and Woods 2002) for edge recognition, convex hull generation, *etc.* They are also used in gray scale images to extract image components. Morphological filters can also be viewed as a superset of the envelope filter. In fact, the envelope filter discussed in the previous section is a basic morphological filtering operation known as *dilation*.

The filter (ball or the line) is referred to as a structuring element, and it acts on the profile to produce different mean lines. One such mean line is the envelope mean line and is produced by dilating the structuring element on the profile.

The process of *dilation* can also be performed on the "underside" of the profile. That is, by rolling a ball below a profile. While not practically feasible, this mean line can be thought of as representing the result of material removal from the surface using the ball or the line structuring element. Such a mean line generated on the bottom is termed the *erosion* mean line.

Dilation and *erosion* are basic morphological operations. There are secondary operations that can be performed on the *dilation* and *erosion* mean lines. For example, the *dilation* envelope can be shifted towards the surface by an amount equal to the radius of the ball (see Exercise 12.4). Instead of simply offsetting the *dilation* envelope towards the surface, another mean line can be generated by eroding the previously generated *dilation* mean line using the same structuring element. Such a mean line, generated by a *dilation* followed by an *erosion* operation, is termed a *closing* mean line. The inverse process, namely the mean line generated by *erosion* followed by *dilation* is an *opening* mean line. *Closing* and *opening* are secondary morphological operations.

Example 12.2. For profile z in Example 12.1, generate the *dilation*, *erosion*, *opening*, and *closing* mean lines using a ball structuring element of size 0.5 mm.

In order to simplify the MATLAB code and make it more efficient, we first write a function for the *dilation* operation. The function accepts two arguments, a profile z and a structuring element B, and outputs the *dilation* mean line (the envelope described in Sect. 12.2). Notice that the profile length has not been increased and that the height computations and comparisons in the beginning and end of the profile have been treated separately (see Exercise 12.1).

```
% Dilation
function C1 = Dilation(z,B)
n = size(z,1);              % length of profile z
m = floor(size(B,1)-1)/2; % length of structuring element B
% First process interior regions of the profile
for j=m+1:n-m
    % compare sum of profile and structuring element
    C1(j,1) = max([z(j-m:j+m,1)+B]);
end
% Process the first m points. If the structuring element has an
% element outside the profile, ignore it. Only consider the
% region where there is overlap
for j=1:m
    C1(j,1) = max([z(1:j+m,1)+B(m-j+2:2*m+1,1)]);
end
% Process the last m points. If the structuring element has an
% element outside the profile, ignore it. Only consider the
% region where there is overlap
for j=n-m+1:n
    C1(j,1) = max([z(j-m:n,1)+B(1:m+(n-j)+1,1)]);
end
```

A similar function can also be written for the *erosion* mean line. However, realizing that the *erosion* mean line is simply the *dilation* mean line computed on the inverse profile, a more elegant function can be written as shown below. *Closing* and *opening* mean lines can be easily computed once *dilation* and *erosion* mean lines are available. Figure 12.3 shows the plots.

```
function C2 = Erosion(z,B)
C2 = -Dilation(-z,B);

function C3 = Closing(z,B)
C3 = Erosion(Dilation(z,B),B);
```

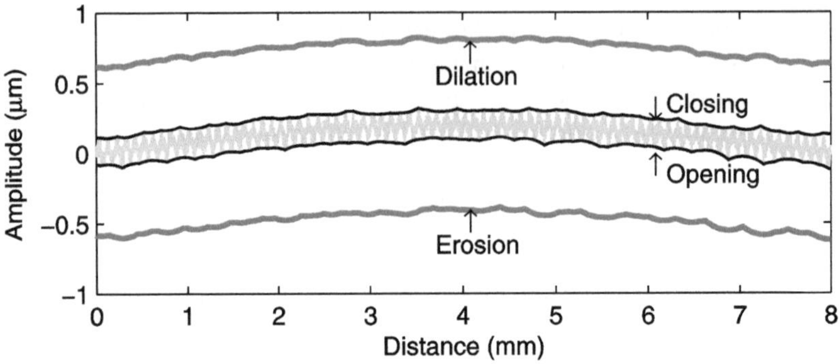

Fig. 12.3 *Dilation*, *erosion*, *opening* and *closing* mean lines for 0.5 mm ball radius

```
function C4 = Opening(z,B)
C4 = Dilation(Erosion(z,B),B);

% Find dilation, erosion, opening and closing mean lines
C1 = Dilation(z,B);
C2 = Erosion(z,B);
C3 = Closing(z,B);
C4 = Opening(z,B);
plot(x,z,'g',x,C1,'r',x,C2,'b',x,C3,'c',x,C4,'m');
xlabel('Distance (mm)');
ylabel('Amplitude (\mum)');
```

12.4 A Word on the Fundamentals

While we have discussed implementations of morphological filters by simulating sliding structural elements on surfaces, there is a deeper underlying fundamental basis for the mathematics in set theory where a set stores the shapes of objects. *Dilation*, *erosion*, and other morphological operations are derived from basic addition and subtraction of sets, also referred to as Minkowski addition and subtraction. Any standard image-processing text describes such operations; see Gonzales and Woods (2002).

12.5 3D Morphological Filtering

We can extend the ideas discussed in Sect. 12.1 to determine the mean surface by simulating a sphere rolling over a surface. Tholath and Radhakrishnan (1999) describe methods to reduce computational time in determining a rolling ball envelope over a 3D surface. In the following examples, we generate a 3D surface and determine the *opening* mean surface by simulating a rolling ball over the surface.

Example 12.3. Generate a 3D sinusoidal surface that contains a uni-directional sine wave of amplitude 1 μm and wavelength of 64 μm. Assume the surface contains 128×128 points at a sampling interval of 1 μm along x and y.

The surface can be generated as shown below.

```
dx = 0.001;                 % spacing in x in mm
dy = 0.001;                 % spacing in y in mm
nx = 128;                   % number of points along x
ny = 128;                   % number of points along y
x = (0:1:nx-1)*dx;          % generate x array
y = (0:1:ny-1)*dy;          % generate y array
for j = 1:ny
    for i = 1:nx
        z(j,i) = 2*sin(2*pi*x(i)/0.064);
    end
end
z = z - mean(mean(z)); % shift to zero mean
mesh(x,y,z);
xlabel('x Distance (mm)');
ylabel('y Distance (mm)');
zlabel('Amplitude (\mum)');
```

Example 12.4. Generate a sphere of radius 16 μm, assuming a sampling interval of 1 μm in x and z.

The sphere can be generated as shown below.

```
Radius = 0.016; % radius in mm
mx = Radius/dx; % number of points in one radius of the ball
my = Radius/dy;
j = 1;
for yr = -Radius:dy:Radius
    i = 1;
    for xr = -Radius:dx:Radius
        if sqrt(xr^2+yr^2) <= Radius
            B(j,i) = sqrt(Radius^2-xr^2-yr^2);
        else
            B(j,i) = 0;
        end
        i = i + 1;
    end
    j = j + 1;
end
```

Example 12.5. Determine the locus of the center of this sphere as it rolls over the surface in Example 12.3.

The locus of the center of the sphere can be obtained as shown below.

```
z1(ny+2*my+1,nx+2*mx+1) =0;
z1(my+1:ny+my,mx+1:nx+mx) = z;
for j=my+1:ny+my
    for k = mx+1:nx+mx
```

```
        C(j-my,k-mx) = max(max(z1(j-my:j+my,k-mx:k+mx)+B));
        % compare sum of profile and ball heights
    end
end
mesh(z); hold on; mesh(C);
xlabel('x Distance (mm)');
ylabel('y Distance (mm)');
zlabel('Amplitude (\mum)');
```

12.6 Summary

Envelope and morphological filters are different from traditional filters. Unlike the premise that different surface bandwidths are responsible for different functional behavior, the underlying assumption here is that surface features such as peaks and valleys influence part functionality, see, for example, Malburg (2003) and Muralikrishnan (2005). A mean line that mimics actual contact mating behavior is therefore determined and used for analysis. This method, although dating back to the 1970s, is still under research and ISO Standards have only recently incorporated these tools. ASME B46.1 does not include these filters. Fast implementations for line structuring elements have been reported (Sedaaghi 1997 and Wang 1994) that significantly improve computational speed.

Exercises

12.1. The code segment shown in Example 12.1 is not computationally efficient. Rewrite it so that the profile length remains the same (no zero-padding in beginning and end), but so that the height computations and comparisons (finding the maximum) are performed only in the region where the ball overlaps the profile. Also, the structuring element is generally symmetric. Therefore, rewrite the program so that only one half of the structuring element is stored in array B.

12.2. Assume profile z in Example 12.1 represents a true surface that is then measured using a surface profile instrument whose stylus radius is 10 μm. Plot the profile as measured by the instrument, and also plot the information that is lost in the digitization process. Research if there are any techniques in the literature to reconstruct the original profile from the raw data (this process is referred to as deconvolution)

12.3. If profile z in Example 12.1 were filtered first using a ball of size 50 μm to generate an envelope mean line (C_2) and this envelope mean line were then filtered using another ball of size 0.5 mm (C_3), will C_3 be identical to the mean line C_1 generated in Example 12.1 directly from profile z? Is there any advantage in using a smaller ball first before using the larger ball for filtering the profile?

12.4. In Example 12.1, notice that the locus of the center of the ball is away from the surface by a distance equal to the radius of the ball. Correct for this offset (by subtracting the radius from the mean lines) to produce mean lines that ride the peaks. Compare this mean line with the *closing* profile in Example 12.2. Suggest applications where each of these mean lines may be used.

References

Gonzales, R.C. and Woods, R.E. 2002, *Digital Image Processing*, 2nd edn, Prentice Hall. Upper Saddle River, NJ, USA.

International Organization for Standardization 2006a, *ISO/TS 16610-40:2006, Geometrical Product Specifications (GPS) – Filtration – Part 40: Morphological Profile Filters: Basic Concepts*, ISO. Geneva, Switzerland.

International Organization for Standardization 2006b, *ISO/TS 16610-41:2006, Geometrical Product Specifications (GPS) – Filtration – Part 41: Morphological Profile Filters: Disk and Horizontal Line-Segment Filters*, ISO. Geneva, Switzerland.

Malburg, M. 2003, 'Surface profile analysis for conformable interfaces', *Transactions of the ASME: Journal of Manufacturing Science and Engineering*, vol. 125, no. 3, pp. 624–627.

Muralikrishnan, B. and Raja, J. 2005, 'Functional filtering and performance correlation of plateau honed surface profiles', *Transactions of the ASME: Journal of Manufacturing Science and Engineering*, vol. 127, no. 1, pp. 193–197.

Radhakrishnan, V. 1972, 'Selection of an enveloping circle radius for E-system roughness measurement', *International Journal of Machine Tool Design and Research*, vol. 12, pp. 151–159.

Radhakrishnan, V. and Von Weingraber, H. 1969, 'Die Analyse digitalisierter Oberflächenprofile nach dem E-System', *Fachberichte für Oberflächentechnik*, vol. 7, no. 11/12, pp. 215–223.

Radhakrishnan, V. and Von Weingraber, H. 1971, 'Zur geometrischen Analyse von Oberflächenprofilen', *Werkstatt und Betrieb*, vol. 104, pp. 239–244.

Scott, P.J. 2000, 'Scale-space techniques', *Proceedings of the X International Colloquium on Surfaces, Chemnitz University of Technology. Chemnitz*, pp. 153–161.

Sedaaghi, M.H. 1997, 'Direct implementation of open-closing in morphological filtering', *Electronic Letters*, vol. 33, no. 3, pp. 198–199

Shunmugam, M. S. and Radhakrishnan, V. 1974, 'Computation of the three-dimensional envelope for roughness measurement', *International Journal of Machine Tool Design and Research*, vol. 14, pp. 211–216.

Srinivasan, V. 1998, 'Discrete morphological filters for metrology', *Proceedings 6th ISMQC Symposium on Metrology for Quality Control in Production, TU Wien, Austria*, September, 1998.

Tholath, J. and Radhakrishnan, V. 1999, 'Three-dimensional filtering of engineering surfaces using envelope system', Precision Engineering, vol. 23, no. 4, pp. 221–228.

Villarrubia, J.S. 1996, 'Scanned probe microscope tip characterization without calibrated tip characterizers', *Journal of Vacuum Science and Technology B: Microelectronics and Nanometer Structures*, vol. 14, no. 2, pp. 1518–1521.

Von Weingraber, H. 1956, 'Zur Definition der Oberflächenrauheit Werkstattstechnik', *Masch. Bau*, vol. 46.

Von Weingraber, H. 1957, 'Über die Eignung des Hullprofils als Bezugslinie für Messung der Rauheit', *Microtechnic*, vol. 11, pp. 6–17.

Wang, D. and He, D-C. 1994, 'A fast implementation of 1-D grayscale morphological filters', *IEEE Transactions on Circuits and Systems-II: Analog and Digital Signal Processing*, vol. 41, no. 9, pp. 634–636

Whitehouse, D.J. 1974, 'Theoretical analysis of stylus integration', *Annals of the CIRP*, vol. 23, pp. 181–182.

Chapter 13
Multi-scale Filtering

13.1 Introduction

The primary purpose of surface texture measurement is to understand the relationship between part functionality and surface finish. For this purpose, we have seen how a surface profile is partitioned into different wavelength bands called roughness, waviness, and form, with the underlying assumption that different wavelength regimes impact part functionality in different ways. Ever so often, an application presents itself where engineers feel that specific functional aspects or process diagnostic information is hidden in several fine surface-wavelength bandwidths. There might be a requirement to tolerance and therefore inspect the amplitude or spacing characteristics within these bandwidths, something not feasible using commercially available software packages. A similar question also presents itself for those using morphological filters. Instead of obtaining a mean line using a rolling ball of a specified size, certain situations may demand more stringent control of features within several specific size ranges. In other words, once again, there is a need for multiple but fine bandwidths.

Filters that have the capability of partitioning a profile into such multiple yet fine bandwidths are known as multi-scale filters, and they are the focus of this chapter. It does not matter whether the bandwidths are defined in terms of surface wavelengths, as in the case of traditional filters, or feature size, as in the case of morphological filters. Because Chap. 12 was on morphological filters, we will continue the thread here and discuss multi-scale filtering using *opening* and *closing* filters. Later, we will discuss why the Gaussian filter is unsuitable as a multi-scale filter and suggest the possibility of using wavelet based filters.

13.2 Alternate Sequence Filters or Scale Space Analysis

When a profile is filtered using a *closing* filter of a certain size, the resulting output does not have any valleys of width narrower than the size of the structuring element.

Similarly, the *opening* filter produces a profile that does not contain any peaks of width narrower than the size of the structuring element. A combination of *opening* and *closing* filters of a certain size will remove all features of width narrower than that of the structuring element.

If a profile is filtered using a combination of *opening* and *closing* filter pairs of varying sizes, the resulting series of profiles will therefore have features in different size ranges. This type of analysis could potentially be useful in understanding the constituents of a profile, not in terms of frequency content, but in terms of feature size content.

This alternating sequence of *opening* and *closing* filter pairs applied to a profile is termed "scale space" analysis. Its application in metrology was reported by Scott (2000) and is now incorporated into ISO standards (ISO/TC 16610-49: 2006) (International Organization for Standardization 2006b). It is performed as shown below.

A profile is first filtered using a *closing-opening* filter pair with a structuring element whose size is approximately equal to the smallest feature size discernible in the profile (typically just larger than the sampling interval or stylus size). The resulting mean line is termed an approximation profile, because it suppresses the details. The difference between this mean line and the profile is termed the difference profile. The difference profile contains features of width narrower than the size of the structuring element, while the approximation profile contains features wider than that of the structuring element.

The approximation profile is then filtered using a structuring element of slightly larger size to yield new approximation and difference profiles. This process can be extended for many levels; the choice of the structuring element sizes are arbitrary, and depend on the application. A potential application of this method is in the identification and removal of outliers in surface profiles. This is illustrated in the following examples.

Example 13.1. Filter profile z in Example 11.1, Chap. 11 using a series of *opening* and *closing* filter pairs of the following structuring element sizes: 10 μm, 100 μm, 1 mm, and 2 mm. Use a line-structuring element in this example. Plot the approximation and difference profiles at each level.

The raw profile z is generated as shown below.

```
n = 8000;               % number of profile points
dx = 0.001;             % spacing in mm
x = (0:1:n-1)'*dx;      % generate x axis
z = detrend(1*sin(2*pi.*x/16)+normrnd(0,0.1*ones(n,1)));
                        % generate profile z
z(2001:2100,1) = -3; % superimpose valley
z(3001:3100,1) = -3; % superimpose valley
```

Then, a line-structuring element is generated as shown below. The width of the line is 10 μm. This line is used to generate the approximation profile at level 1. The approximation profile is simply the sequence of *opening* and *closing* mean lines generated on profile z. The approximation at level 1 does not contain any features narrower than 10 μm. The difference profile at level 1 is the difference between the original profile and the approximation at level 1. Obviously, the difference profile at this level only contains features narrower than the size of the structuring element (10 μm).

```
% Generate line of size 10 um and filter
Radius = 0.005;                      % half width of line
m = Radius/dx;                       % number of points in half width
xr = (-Radius:0.001:Radius)';        % generate x array
B = ones(size(xr,1),1);              % generate structuring element
Approx_1 = Opening(Closing(z,B),B); % determine Approximation
                        % profile = opening + closing combination
Diff_1 = z - Approx_1;               % determine Difference profile
```

We next generate the larger structuring element, the line of width 0.1 mm. We use the approximation profile from the previous level as the raw profile and generate the new approximation and difference profiles at this level.

```
% Generate line of size 100 um and filter
Radius = 0.05;                    % half width of line
m = Radius/dx;                    % number of points in half width
xr = (-Radius:0.001:Radius)'; % generate x array
B = ones(size(xr,1),1);           % generate structuring element
Approx_2 = Opening(Closing(Approx_1,B),B);
Diff_2 = Approx_1 - Approx_2; % determine Difference profile
```

We repeat the process for the other two structuring elements. The approximation and difference profiles at each level are plotted in Fig. 13.1. Recall that the width of the valleys was 100 μm. Because these valleys have greater width than the structuring element at level 1 (10 μm width), the approximation profile at level 1 captures the valleys. However, the difference profile at level 2 retains the valleys because the structuring element at level 2 is of equal size to the valley width.

```
% Generate line of size 1 mm and filter
Radius = 0.5;
m = Radius/dx;
xr = (-Radius:0.001:Radius)';
```

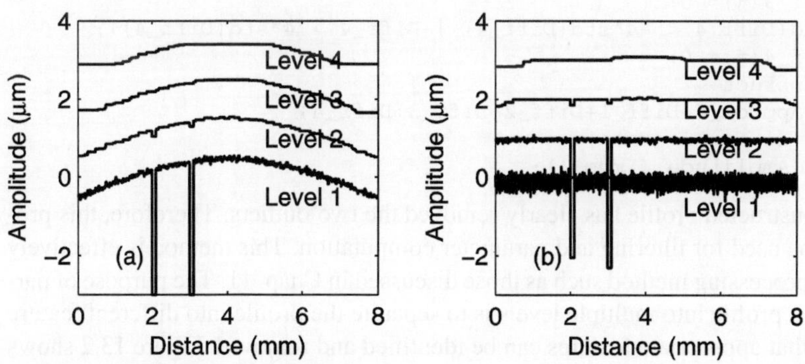

Fig. 13.1 **a** Approximation and **b** difference profiles

```
B = ones(size(xr,1),1);
Approx_3 = Opening(Closing(Approx_2,B),B);
Diff_3 = Approx_2 - Approx_3;

% Generate line of size 2 mm and filter
Radius = 1;
m = Radius/dx;
xr = (-Radius:0.001:Radius)';
B = ones(size(xr,1),1);
Approx_4 = Opening(Closing(Approx_3,B),B);
Diff_4 = Approx_3 - Approx_4;

plot(x,Approx_1,x,Approx_2+1,x,Approx_3+2,x,Approx_4+3);
xlabel('Distance (mm)');
ylabel('Amplitude (\mum)');
plot(x,Diff_1,x,Diff_2+1,x,Diff_3+2,x,Diff_4+3);
xlabel('Distance (mm)');
ylabel('Amplitude (\mum)');
```

Example 13.2. Reconstruct the outlier removed profile in the following manner. Use a global threshold of four standard deviations on all difference profiles in the preceding example. That is, for each difference profile, if a profile element were outside of the four standard deviation band from the mean, set that profile element value to zero. Then, add all difference profiles and the approximation profile from the last filtering operation (2 mm structuring element size) to reconstruct the outlier removed profile.

Threshold the difference profiles using the four standard deviation condition specified in the problem. Then, compute the sum of all difference profiles and the approximation from the last filtering operation to plot the reconstructed profile.

```
% Threshold the difference profiles
j = find(Diff_1 < -4*std(Diff_1) | Diff_1 > 4*std(Diff_1));
    Diff_1(j)= 0;
j = find(Diff_2 < -4*std(Diff_2) | Diff_2 > 4*std(Diff_2));
    Diff_2(j)= 0;
j = find(Diff_3 < -4*std(Diff_3) | Diff_3 > 4*std(Diff_3));
    Diff_3(j)= 0;
j = find(Diff_4 < -4*std(Diff_4) | Diff_4 > 4*std(Diff_4));
    Diff_4(j)= 0;
% Reconstruct
plot(x,Approx_4+Diff_1+Diff_2+Diff_3+Diff_4)
xlabel('Distance (mm)');
ylabel('Amplitude (\mum)');
```

The reconstructed profile has clearly removed the two outliers. Therefore, this profile can be used for filtering and parameter computation. This method is effectively a robust processing method such as those discussed in Chap. 11. The purpose of partitioning a profile into multiple levels is to separate the profile into different feature sizes so that appropriate features can be identified and removed. Figure 13.2 shows the original and reconstructed profiles.

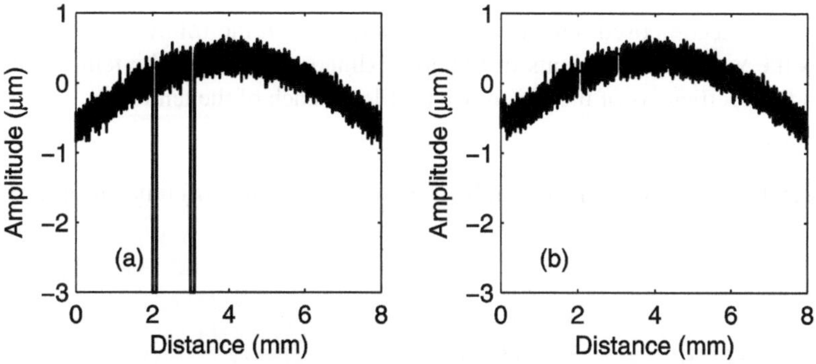

Fig. 13.2 **a** Profile z and **b** outlier-removed profile

13.3 Wavelet-Based Filters

Wavelet-based filtering and its application in surface metrology is an active topic of research (Chen *et al.*, 1995; Jiang *et al.*, 2000; Fu *et al.*, 2003). Here, we provide a very brief overview of how wavelet-based filters can be used in surface profile filtering.

In a nutshell, wavelet-based filtering can be regarded as the E system equivalent of the alternate sequence filters of the M system. That is, the result of a wavelet-based filter will be a set of several filtered profiles, each corresponding to a certain wavelength bandwidth. The traditional Gaussian filter does not have steep transmission characteristics and therefore it is not possible to obtain close partitioning of the profile into multiple-wavelength bandwidths. Some wavelets, on the other hand, provide steep transmission characteristics and enable partition of profiles into narrow bandwidths.

A wavelet family has two filters associated with it: a scaling function that can be regarded as a low-pass filter and a wavelet filter that can be regarded as a bandpass filter. The transmission characteristics of these filters therefore determine if they are an appropriate choice for surface texture analysis; Fu *et al.* (2003) recommend Bior6.8 and Coif4 on the basis of the transmission characteristics of the corresponding filters. The international Standards describe spline wavelets (International Organization for Standardization 2006a). In the following example, we show how a wavelet-based filter can be quickly implemented using the Wavelet toolbox in MATLAB.

Example 13.3. Generate a sinusoidal profile 8 mm long with 1 μm spacing with the following components: a sine wave of amplitude 0.5 μm and wavelength 16 mm, a sine wave of amplitude 0.2 μm with wavelength of 4 mm and normally distributed random noise of zero mean and standard deviation 0.1 μm. Decompose the profile into ten levels using Coif4 wavelet and plot the difference profiles and the approximation profile.

The profile z can be generated as shown below:

```
x = (0:1:7999)'*0.001;
z = 0.5*sin(2*pi.*x/16) + 0.2*sin(2*pi.*x/4) +
normrnd(0,0.1*ones(8000,1));
```

Profile z can be decomposed into wavelet coefficients C in the following manner: C contains the wavelet coefficients of the approximation profile at the tenth level along with the coefficients of the difference profiles at each of the ten levels.

```
[C,L] = wavedec(z,10,'coif4');
```

At any given level, the coefficients can be used to recreate the difference profile as shown.

```
n=8000;step=0;gdelta=0.5;
for i=1:10
    hold on
    D = wrcoef('d',C,L,'coif4',i);
    plot(x,D+step);
    text(x(n),step,['Diff' num2str(i)]);
    step = step + gdelta;
end
hold on
```

The approximation profile at the tenth level can be recreated as shown.

```
A = wrcoef('a',C,L,'coif4',10);
plot(x,A+step,'g');
text(x(n),step,['Approx' num2str(10)]);
step = step + gdelta;
plot(x,z+step,'r');
text(x(n),step,'Profile');
xlabel('Distance (mm)');
ylabel('Amplitude (\mum)');
title('Multi Resolution Analysis');
axis([0 8 -0.5 7])
```

We have shown the different difference profiles stacked one above the other in Fig. 13.3, which is why the amplitude axis scales up to about 7 μm. It should be

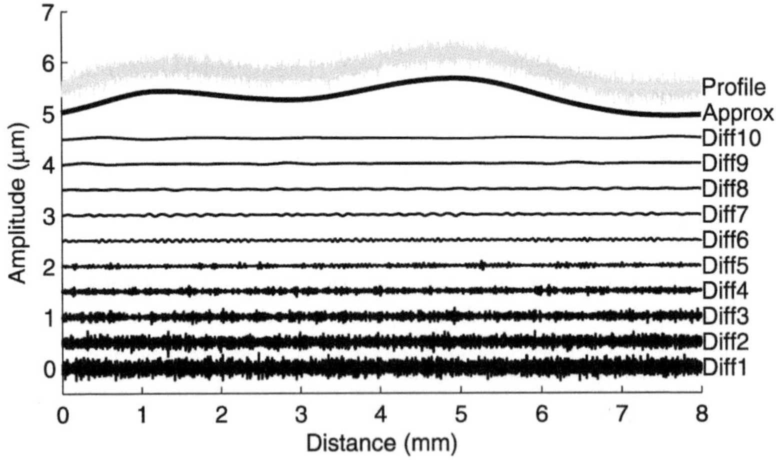

Fig. 13.3 Profile z, difference profiles and the approximation profile

noted that the different levels correspond to different surface-wavelength bands. The difference profile at level 1 represents a bandwidth between two and four times the spacing. The difference profile at level 2 represents four to eight times the spacing. The next level represents an interval of twice the previous. Level 10 therefore represents an interval of 1,024 to 2,048 times the spacing. The approximation at level 10 represents all wavelengths greater 2,048 times the spacing.

13.4 Summary

Alternate sequence filters and wavelet based multi-resolution analysis can be regarded as logical extensions of the corresponding rolling ball and Gaussian type filters to partition a profile into fine bandwidths. Wavelet based filtering is a very active research topic and we have barely even scratched the tip of the iceberg here. We urge the reader to review several excellent texts on this topic (Hubbard 1998; Mallat 1999; Strang and Nguyen 1996).

In this part (Part II), we have traced the evolution of several new filtering techniques many of which are being incorporated into international Standards. As a consequence, these filters that are currently primarily in research phase will soon find widespread industrial use. In Table 13.1 below, we summarize the key attributes of these filters.

Table 13.1 Advantages and limitations of different filtering techniques

Filter	Advantages	Limitations
2RC	• Earliest filter used in surface metrology • Implemented first in hardware • Subsequent digital representation	• Phase distortion in the output is a major limitation • Need separate roughness and waviness filters • Edge distortion • Cannot handle large form • Not robust against outliers
Gaussian	• Widely used today • No phase distortion, which is a key attribute • Single filter can be used to obtain both roughness and waviness	• Edge distortion • Cannot handle large form • Not robust against outliers
Zero-order Gaussian regression	• Overcomes edge distortions • Advantageous when measuring short traces where edge data cannot be discarded	• Cannot handle large form • Not robust against outliers
Second-order Gaussian regression	• Overcomes edge distortions • Handles large form	• Not robust against outliers • Computationally intensive

Table 13.1 (continued)

Filter	Advantages	Limitations
Robust Gaussian regression	• Robust against outliers • Advantageous when measuring surfaces with deep scratches (machining cast iron, for instance)	• Computationally intensive because multiple passes are necessary
Spline	• Overcomes edge distortions • Handles large form • Computationally efficient • Sharp transmission in comparison to Gaussian	• Not robust against outliers • No weighting function expression • Cumbersome implementation
Robust spline (draft international Standard)	• Overcomes edge distortions • Handles large form • Robust against outliers	• No weighting function expression • Cumbersome implementation
Rk (Gaussian)	• Robust against outliers	• Edge distortion • Multiple passes may be necessary
Rk (Spline)	• Robust against outliers • No edge distortion	• Multiple passes may be necessary
Morphological, scale-space	• Applications such as outlier removal, mating surface, rolling or sliding surface characterization • Computationally efficient algorithms are available for line-structuring elements	• Not yet widely popular in traditional metal cutting industry • Commercial instrumentation software rarely incorporate these tools
Wavelets	• Sharp transmission characteristics • Finer bands imply possibility of demonstrating link to function	• Computationally intensive • Topic under research

Exercises

13.1. For any given structuring element size, if the *opening* operation were performed prior to the *closing* operation in the alternate sequence filter, will the approximation and difference profiles be different from that in Example 13.1.

13.2. A scheme similar to the alternate sequence filter can conceivably be used to produce a series of approximation and difference profiles in traditional filtering also, say, using a Gaussian filter with the series of cutoffs: 8 μm, 80 μm, 0.8 mm, 8 mm, *etc*. The waviness profile in each filtering operation can be termed the approximation profile and the roughness can be termed the difference profile. Explain why the Gaussian filter is not an ideal choice for such bandwidth partitioning? Describe characteristics of a filter that can be used for such partitioning?

13.3. How would the alternate sequence be performed in reverse order of structuring element sizes? If a profile is first filtered using a large structuring element to produce an approximation and a difference profile, how will the approximation and difference profiles for the next lower structuring element size be obtained?

13.4. Generate a sinusoidal profile of amplitude 1 μm and wavelength 0.384 μm (spacing 1 μm and profile length 8 mm). Decompose this profile using the Coif4 wavelet at ten levels and plot the difference profiles as in Fig. 13.3. At what level do you see expect to see the sinusoid?

References

Chen, X., Raja, J. and Simanapalli, S. 1995, 'Multi-scale analysis of engineering surfaces', *International Journal of Machine Tools and Manufacture*, vol. 35, no. 2, pp. 231–238.

Fu, S., Muralikrishnan, B. and Raja, J. 2003, 'Engineering surface analysis with different wavelet bases', *Transactions of the ASME: Journal of Manufacturing Science and Engineering*, vol. 125, no. 4, pp. 844–852.

Hubbard, B.B. 1998, *The World According To Wavelets: The Story of a Mathematical Technique in the Making*, 2nd edn, A K Peters. Wellesley, MA, USA.

International Organization for Standardization 2006a, *ISO/TS 16610-29:2006, Geometrical Product Specifications (GPS) – Filtration – Part 29: Linear Profile Filters: Spline Wavelets*, ISO. Geneva, Switzerland.

International Organization for Standardization 2006b, *ISO/TC 16610-49:2006, Geometrical Product Specifications (GPS) – Filtration – Part 49: Morphological Profile Filters: Scale Space Techniques*, ISO. Geneva, Switzerland.

Jiang, X. Q., Blunt, L. and Stout, K. J. 2001, 'Development of a lifting wavelet representation for surface characterization', *Proceedings of the Royal Society of London Series A Mathematical Physical and Engineering Sciences*, vol. 456, pp. 2283–2313.

Mallat, S. 1999, *A Wavelet Tour of Signal Processing*, 2nd edn, Academic Press. San Diego, CA, USA.

Scott, P.J. 2000, 'Scale-space techniques', *Proceedings of the X International Colloquium on Surfaces, Chemnitz University of Technology. Chemnitz*, pp. 153–161.

Strang, G. and Nguyen, T. Q. 1996, *Wavelets and Filter Banks*, 2nd edn, Wellesley-Cambridge Press. Wellesley, MA, USA.

Part III
Fitting

Part III is devoted to selected topics on fitting substitute geometry, as applied to surface and roundness metrology. The purpose of determining best-fit geometry is to produce perfect models of otherwise imperfect parts. The models can then be used to extract information such as dimension, form, and maybe even position. Fitting is used extensively in coordinate metrology where it is a necessary analysis step. We provide a brief overview of the fitting process in Chap. 14. In Chap. 15, we discuss the problem of fitting a 2D line to a set of points. We use matrix methods, extend the problem to a line in 3D and also mention a solution strategy for best-fit planes. In Chap. 16 we discuss iterative algorithms that are applicable to more complex geometries, and extend this discussion for circles, spheres and cylinders in Chap. 17. In Chap. 18, we discuss a linear approximation for the circle; a formulation that allows us to easily analyze roundness data acquired from a rotating spindle instrument. We discuss exchange algorithms in Chap. 19 for the determination of minimum-zone boundaries for lines, planes, and circles. In Chap. 20, we illustrate the use of a linear programming technique for determining the minimum-zone circle and then show it is in fact linked to exchange algorithms discussed previously.

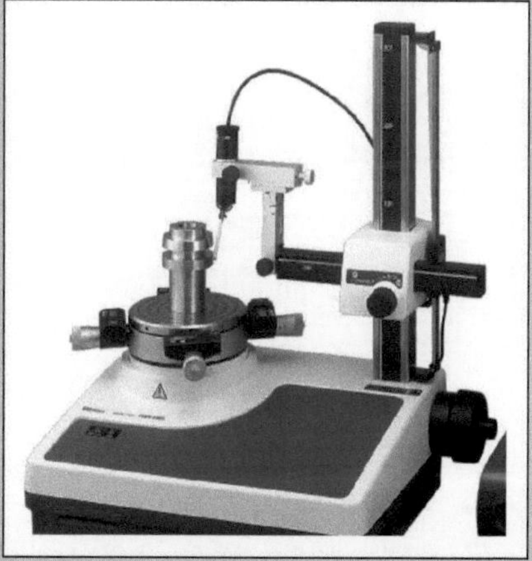

Fig. III Roundness measurement using a rotating spindle roundness instrument (courtesy: Mitutoyo)

Chapter 14
Introduction to Fitting Substitute Geometry

14.1 Introduction

Fitting best-fit geometry is the process of establishing substitute geometry from real data collected using metrology instruments on real components. It is a necessary analysis step in coordinate metrology because data obtained from a coordinate measuring machine (CMM) are essentially point coordinates on the surface of the part, not values of the features of interest such as diameter, form, length, height, straightness, parallelism, *etc.* (Bosch 1995; Srinivasan 2007).

In order to transform the set of point coordinates to estimates of the features of interest, we fit relevant substitute geometry to the point data set and create a model of the part we are measuring. The model then provides information that can be used for process control, functional analysis, tolerance verification, *etc.* In the simple example of determining the diameter of a cylinder, a CMM can be programmed to acquire coordinates of many points on the surface. A best-fit cylinder fit to this data provides information on the diameter and form of the part. Determining the parameters of a best-fit geometry is a very well-researched topic (see Sect. 14.3 for a sampling of literature in this field). It is clear that best-fit geometry is created for every geometric element that is measured on a CMM. Examples of such geometries include circles, cylinders, spheres, planes, lines, cones, tori, *etc.*

In roundness metrology, the primary aim of a measurement is to evaluate the extent of roundness, cylindricity, and sometimes even straightness of a part. Recall that a roundness instrument only measures deviations. However, when mounting a part on a rotating-spindle instrument, it is rarely possible to center the part precisely. In measuring cylindricity, the part is likely to have some tilt and eccentricity. Therefore, the deviations as measured do not directly provide the desired out-of-roundness or out-of-cylindricity. The out-of-roundness (or out-of-cylindricity) of a component is evaluated from a reference circle (or cylinder), which is determined using a fitting process. The center of the reference circle provides a measure of eccentricity. The center and direction cosines of the axis of the reference cylinder provide information on eccentricity and tilt for cylinders.

Fitting is not as commonly used in surface metrology as a necessary analysis step. However, in many situations, it is an important preprocessing step. That is, prior to filtering, the nominal geometry of the part is often best-fit and removed from the surface texture data. For example, it is difficult to align the part parallel to the internal datum of the instrument. Therefore, an essential preprocessing step involves removing any residual tilt after alignment by fitting lines (in the case of profiles) or planes (in the case of 3D surfaces). When measuring surface profiles or area maps on curved surfaces such as a sphere or a cylinder, the data is generally preprocessed by best fitting and removing the nominal shape of the surface. This is to diminish the influence of the nominal geometry on the filtering process.

14.2 Fitting Criteria

If there are exactly as many data points collected as are required to determine the parameters of geometry, then fitting criteria has no relevance. While three points are adequate to determine the diameter of a circle in a plane, we sample more points to reduce the influence of outliers, to capture part form errors, to minimize the effect of any measurement errors, *etc*. In such an over-determined system where we have more data points than unknown parameters, fitting criteria play an important role in the parameter determination process.

Perhaps the most commonly used criteria to determine parameters of best-fit geometry is to minimize the sum of squared deviations, also known as the least-squares (LS) method. LS technique is robust against outliers and produces a stable best-fit surface. LS best-fit estimates of some simple geometries can be determined quite easily by solving a set of simultaneous equations. Estimates for more complex geometries can be determined using iterative techniques.

While the LS estimate is robust against outliers and relatively easy to estimate, the minimum-zone (MZ) criteria is the basis in determining tolerance compliance. In fact, ASME Y14.5M Standards (American Society of Mechanical Engineers 1994a, 1994b) describe tolerance boundaries under MZ criteria. The Standards however do not indicate a method for the determination of the zone, providing engineers and mathematicians with the challenging problem of determining value of the zone width for a variety of geometries.

In the case of closed geometries such as circles and cylinders, minimum-circumscribed (MC) and maximum-inscribed (MI) criteria are also popular. All four criteria, LS, MZ, MI, and MC, are mentioned in American (American Society of Mechanical Engineers 1972) and international Standards (International Organization for Standardization 1985) that describe roundness. MZ fitting is rarely used in surface texture analysis; it does not fall under the purview of the ASME B46.1 or corresponding ISO Standards. However, form measurements (such as straightness or flatness) are usually reported using the MZ method. An example of LS, MC, MI, and MZ boundaries are shown for a roundness profile in Fig. 14.1.

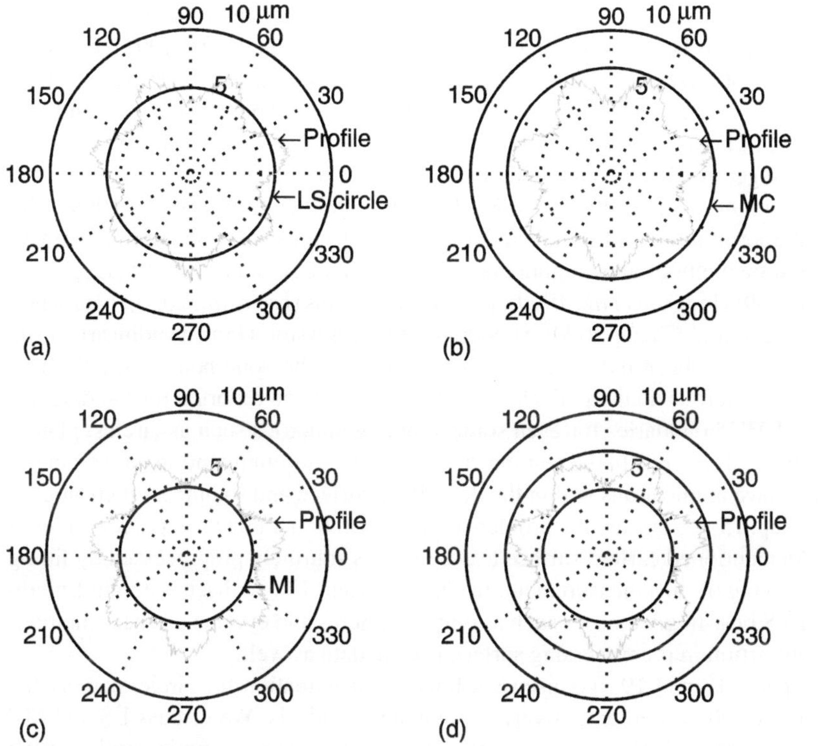

Fig. 14.1 **a** Least-squares (LS) circle. **b** Minimum-circumscribed (MC) circle. **c** Maximum-inscribed (MI) circle. **d** Minimum-zone (MZ) circle

14.3 Solution Methodologies

The determination of LS, MZ, MI, and MC boundaries for any geometry (MI and MC for closed geometries only) is an optimization problem and can be solved using constrained and/or unconstrained optimization techniques. Known solution methodologies exist for different LS geometries (Forbes 1989; Shakarji 1998). Determining MI, MC, and MZ boundaries for different geometries is a challenging task and an active topic of research.

The determination of MZ, MI, and MC boundaries are simplified if the objective function (a function that is minimized to obtain the parameters of the desired geometry) and constraints are linear in the unknown parameters. Luckily, in surface and roundness metrology, we generally encounter simple geometries that either are intrinsically linear, or can be approximated by linear forms. In such cases, exchange algorithms that use simple geometry based computations are the preferred approach. We discuss exchange algorithms for lines, planes, and circles in Chap. 19.

For the general case of determining MZ geometry for CMM data, the methods reported in the literature are diverse. We present here only a brief sampling of the literature to illustrate the variety of techniques that have been reported; a comprehensive review is beyond the scope of this text. See Feng and Hopp (1991) for a short summary. Optimization techniques such as down-hill simplex or other methods have been used for determining MZ boundaries (Murthy and Abdin 1980; Carr and Ferreira 1995; Cheraghi *et al.*, 1996; Kanada 1996). Some researchers have used computational geometry-based methods such as convex hull and Voronoi diagrams (Lai and Wang 1988; Traband *et al.*, 1989; Etesami and Qiao 1990; Huang 1999). Numerical approximation of the objective function (Dhanish and Shunmugam 1991), enumeration methods and combinatorics (Ventura and Yeralan 1989; Rajagopal and Anand 1999; Hodgson *et al.*, 1999), genetic algorithms (Lai *et al.*, 2000), neural networks (Suen and Chang 2000), support vector regression (Prakasvudhisarn *et al.*, 2003), *etc.* have been used for the determination of the zone boundaries. Clearly, there are numerous mathematical methods that have been reported for the determination of MZ boundaries for even some simple geometries such as circles, planes, cylinders, and spheres. In the coming chapters, we focus our attention only on methods that have applications within the domain of surface and roundness metrology.

In Chap. 15, we discuss the problem of fitting LS best-fit line and plane to a set of data. As mentioned earlier, surface texture data is routinely preprocessed by fitting and removing tilt in a measurement. In Chaps. 16 and 17, we discuss iterative methods for LS best-fit geometries such as circles, spheres and cylinders. These methods find application in preprocessing surface texture data as well.

In Chaps. 18 and 19, we discuss a linear approximation for circle, namely the limaçon, which is used extensively in roundness analysis. We discuss LS and MZ best-fit boundaries in these chapters. We discuss a linear programming optimization method in Chap. 20 and show how this method may be used to derive MC and MI boundaries for the limaçon. The MI and MC boundaries are used extensively in roundness analysis as well.

References

American Society of Mechanical Engineers 1972, *ASME B89.3.1-1972, Measurement of Out-of-Roundness,* ASME. New York, NY, USA.

American Society of Mechanical Engineers 1994a, *ASME Y14.5M-1994, Dimensioning and Tolerancing,* ASME. New York, NY, USA.

American Society of Mechanical Engineers 1994b, *ASME Y14.5.1M-1994, Mathematical Definition of Dimensioning and Tolerancing Principles*, ASME. New York, NY, USA.

Bosch, R. (ed) 1995, *Coordinate Measuring Machines and Systems*, 1st edn, CRC Press. New York, USA.

Carr, K. and Ferreira, P. 1995, 'Verification of form tolerances part I: Basic issues, flatness, and straightness', *Precision Engineering*, 17, pp. 131–143.

Cheraghi, S.H., Lim, H.S. and Motavalli, S. 1996, 'Straightness and flatness tolerance evaluation: an optimization approach', *Precision Engineering*, vol. 18, pp. 30–37.

Dhanish, P.B. and Shunmugam, M.S. 1991, 'An algorithm for form error evaluation – using the theory of discrete and linear Chebyshev approximation', *Computer Methods in Applied Mechanics and Engineering*, vol. 92, pp. 309–324.

Etesami, F. and Qiao, H. 1990, 'Analysis of two-dimensional measurement data for automated inspection', *Journal of Manufacturing Systems*, vol. 9, pp. 21–34.

Feng, S. and Hopp, T. 1991, *A Review of Current Geometric Tolerancing Theories and Inspection Data Analysis Algorithms*, NISTIR 4509, National Institute of Standards and Technology. Gaithersburg, Maryland.

Forbes, A.B. 1989, *Least-Squares Best-Fit Geometric Elements*, NPL Report DITC 140/89, National Physical Laboratory. Teddington, UK.

Hodgson, T.J., Kay, M.G., Mittal, R.O. and Tang, S. 1999, 'Evaluation of cylindricity using combinatorics', *IIE Transactions*, vol. 31, pp. 39–47.

Huang, J. 1999, 'An exact minimum zone solution for three-dimensional straightness evaluation problems', *Precision Engineering*, vol. 23, pp.204–208.

Kanada, T. 1996, 'Computation of sphericity from minimum circumscribing and maximum inscribing centers by means of simulation data and downhill Simplex method', *International Journal of Japan Society for Precision Engineering*, vol. 30, no. 3, pp. 253–258.

International Organization for Standardization 1985, *ISO 4291:1985, Methods for the Assessment of Departure from Roundness – Measurement of Variations in Radius*, ISO. Geneva, Switzerland.

Lai, H.Y., Jywe, W.Y., Chen, C.K. and Liu, C.H. 2000, 'Precision modeling of form errors for cylindricity evaluation using genetic algorithms', *Precision Engineering*, vol. 24, pp. 310–319.

Lai, K. and Wang, J. 1988, 'Computational geometry approach to geometric tolerancing', *Proceedings of the XVI North American Manufacturing Research Conference*, pp. 376–379. Urban-Campaign, IL, USA.

Murthy, T.S.R. and Abdin, S.Z. 1980, 'Minimum zone evaluation of surfaces', *International Journal of Machine Tools Design and Research*, vol. 20, pp. 123–136.

Prakasvudhisarn, C., Trafalis, T. and Raman, S. 2003, 'Support vector regression for determination of minimum zone', *Transactions of the ASME: Journal of Manufacturing Science and Engineering*, vol. 125, pp. 736–739.

Rajagopal, K. and Anand, S. 1999, 'Assessment of circularity error using a selective data partitioning approach', *International Journal of Production Research*, vol. 37, no. 17, pp. 3959–3979.

Shakarji, C. 1998, 'Least-squares fitting algorithms of the NIST Algorithm Testing System', *Journal of Research of the NIST*, vol. 103, no. 6, pp. 633–641.

Srinivasan, V. 2007, 'Computational metrology for the design and manufacture of product geometry: A classification and synthesis', *Transactions of the ASME: Journal of Computing and Information Science in Engineering*, vol. 7, no. 1, pp. 3–9.

Suen, D.S. and Chang, C.N. 1997, 'Application of neural network interval regression method for minimum zone straightness and flatness', *Precision Engineering*, vol. 20, pp. 196–207.

Traband, M.T., Joshi, S., Wysk, R.A. and Cavalier, T.M. 1989, 'Evaluation of straightness and flatness tolerances using the minimum zone', *Manufacturing Review*, vol. 2, pp. 189–195.

Ventura, J.A. and Yeralan, S. 1989, 'The minimax center estimation problem for automated roundness inspection', *European Journal of Operations Research*, vol. 41, pp. 64–72.

Chapter 15
Least-Squares Best-Fit Line and Plane

15.1 Introduction

Surface profile data (2D profiles or 3D topography) acquired using metrology instruments may possess residual tilt even after careful alignment with the datum within the instrument. Least-squares (LS) best-fit lines and planes are used to remove this tilt prior to filtering during surface texture analysis. We therefore discuss methods for the determination of LS best-fit lines and planes in this chapter. LS best-fit line problems have closed form solutions that are discussed in standard engineering mathematics textbooks (Kreyszig 2005). We briefly review that here, and discuss matrix-based solutions for both lines and planes.

15.2 Closed-Form Solution for LS Best-Fit Line

Let $(x(i), z(i))_{i=1:n}$ be a set of n coordinates of a surface profile [$z(i)$ is the measured height of the ith point and $x(i)$ is the distance of the ith point from the first point]. Let the best-fit line be defined as $z = a + bx$, where a is the intercept and b is the slope. Then, a LS objective function E can be defined as the sum of the squared distances (also referred to as deviation) of each ordinate from the line. Thus, E is given by

$$E = \sum_{i=1}^{n} d(i)^2 , \tag{15.1}$$

where $d(i)$ is the deviation function (deviation or distance of a point from the line) given by

$$d(i) = z(i) - a - bx(i) . \tag{15.2}$$

The parameters of the line can be determined by partially differentiating E with respect to a and b and solving the set of simultaneous equations given in Eq. 15.3:

$$\begin{aligned} \frac{\partial E}{\partial a} &= 2(-1)\sum_{i=1}^{n}(z(i)-a-bx(i))=0 \quad \Rightarrow na+b\sum_{i=1}^{n}x(i)=\sum_{i=1}^{n}z(i) \\ \frac{\partial E}{\partial b} &= 2\sum_{i=1}^{n}(-x(i))(z(i)-a-bx(i))=0 \Rightarrow a\sum_{i=1}^{n}x(i)+b\sum_{i=1}^{n}x(i)^2=\sum_{i=1}^{n}x(i)z(i) \end{aligned} \tag{15.3}$$

Solving Eq. 15.3, we get:

$$\begin{aligned} b &= \frac{\sum_{i=1}^{n}x(i)\sum_{i=1}^{n}z(i)-n\sum_{i=1}^{n}x(i)z(i)}{\left(\sum_{i=1}^{n}x(i)\right)^2-n\sum_{i=1}^{n}x(i)^2} \\ a &= \frac{\sum_{i=1}^{n}x(i)\sum_{i=1}^{n}x(i)z(i)-\sum_{i=1}^{n}x(i)^2\sum_{i=1}^{n}z(i)}{\left(\sum_{i=1}^{n}x(i)\right)^2-n\sum_{i=1}^{n}x(i)^2} \end{aligned} \tag{15.4}$$

The deviation function $d(i)$ is linear in its parameters (a, b), making it convenient to derive the closed-form solution as shown here.

15.3 Matrix Formulation

We present here an alternative formulation to obtain solutions for (a, b) without having to explicitly derive the partial differentials. The advantage here is that this approach lends itself to easier implementation. We exploit the fact that the objective function E, being a sum of squared terms, can only take a value as small as zero when each of its constituent deviation terms are zero. Therefore, equating each deviation term to zero, we have a system of n equations in two variables of the form $z(i) = a + bx(i)$, $i = 1$ to n. This is an over-constrained system and can be written in matrix form $B = AP$, where

$$B = \begin{bmatrix} z(1) \\ z(2) \\ \vdots \\ z(n) \end{bmatrix}, \quad A = \begin{bmatrix} 1 & x(1) \\ 1 & x(2) \\ \vdots & \vdots \\ 1 & x(n) \end{bmatrix}, \quad P = \begin{bmatrix} a \\ b \end{bmatrix}. \tag{15.5}$$

We now use an important result from linear algebra to proceed. Accordingly, *the LS solution for an equation of the form $AP = B$ is the solution of the corresponding normal equation $A^TAP = A^TB$*. Thus, the normal equation in this example is clearly a system of two equations in two variables (A^TA is of size 2×2, A^TB is of size 2×1), which can be readily solved. Standard texts discuss the normal equation representation (Heath 2002; Stoer *et al.*, 2002).

To see why the normal equation rule is valid at all, we proceed backwards. Expanding $A^T AP = A^T B$, we get

$$\begin{bmatrix} 1 & 1 & \dots & 1 \\ x(1) & x(2) & \dots & x(n) \end{bmatrix} \begin{bmatrix} 1 & x(1) \\ 1 & x(2) \\ \vdots & \vdots \\ 1 & x(n) \end{bmatrix} \begin{bmatrix} a \\ b \end{bmatrix} = \begin{bmatrix} 1 & 1 & \dots & 1 \\ x(1) & x(2) & \dots & x(n) \end{bmatrix} \begin{bmatrix} z(1) \\ z(2) \\ \vdots \\ z(n) \end{bmatrix} . \tag{15.6}$$

Simplifying, we obtain

$$\begin{bmatrix} n & \Sigma x(i) \\ \Sigma x(i) & \Sigma x(i)^2 \end{bmatrix} \begin{bmatrix} a \\ b \end{bmatrix} = \begin{bmatrix} \Sigma z(i) \\ \Sigma x(i) z(i) \end{bmatrix} , \tag{15.7}$$

which is the matrix representation of Eq. 15.3. Obtaining the LS solution from the normal form is an important result, one that we will use over and over again in a variety of situations.

15.4 Centroid as a Point on the LS Line

The mathematics of determining the LS estimate can be simplified by making a small observation as shown by Forbes (1989). For any line given by $z = a + bx$, where a and b are defined in Eq. 15.4, it can be seen that the LS line passes through the centroid of the data given by $\left(\sum_{i=1}^{n} x(i)/n, \sum_{i=1}^{n} z(i)/n \right)$. Therefore, if the data were translated so that the centroid is the new origin, then $\sum_{i=1}^{n} x(i) = \sum_{i=1}^{n} z(i) = 0$.

Example 15.1. Show that the average of n observation of a variable x is in fact a LS estimate.

Consider n observations $x(i)$, $i = 1$ to n and let x_a be a best guess estimate. Then, according to the LS method, we determine x_a by minimizing the sum of squared errors of each observation from x_a. That is, we determine x_a by minimizing the objective function E defined as

$$E = \sum_{i=1}^{n} (x(i) - x_a)^2 .$$

Partially differentiating E with respect to x_a and equating to zero, we get

$$\frac{\partial E}{\partial x_a} = 2 \sum_{i=1}^{n} (x(i) - x_a) = 0 \Rightarrow x_a = \frac{\sum_{i=1}^{n} x(i)}{n} .$$

Thus, x_a is indeed the average of the n observations.

Example 15.2. Determine the parameters of a LS best-fit line computed using vertical deviations for the data given by the arrays $x = [1.1\ 2.2\ 3.1\ 4.3\ 5.2]$ and $z = [0.9\ 2.1\ 3.0\ 4.2\ 5.1]$, using the method described in Sect. 15.1.

The following MATLAB segment implements the computations described in Sect. 15.1. The parameters of the best-fit line can also be obtained using MATLAB's inbuilt `polyfit` command.

```
x= [1.1 2.2 3.1 4.3 5.2];
z = [0.9 2.1 3.0 4.2 5.1];
% Method described in section 15.1
n = size(x,2);
b = (sum(x)*sum(z)-n*sum(x.*z))/(sum(x)^2-n*sum(x.^2));
a = (sum(z) - b*sum(x))/n;
[b a]
% Using MATLAB's polyfit command
polyfit(x,z,1)
```

Example 15.3. For the data set given in Example 15.2, generate the parameters of the best-fit line using the matrix method described in Sect. 15.2.

The following MATLAB segment generates array B and matrix A, and computes the parameters using the inbuilt function \ (type help `slash` at the MATLAB prompt). Notice that the matrices $A^T AP$ and $A^T B$ are not explicitly evaluated. The left-division function automatically computes the LS estimate using the normal form of $AP = B$.

```
B = [z'];
A = [ones(n,1) x'];
P = A\B;
[P(2) P(1)]
```

15.5 Normal LS in Parametric Form

In the discussion thus far, we have considered the line in its slope-intercept form. Such a formulation is obviously not scalable to a line in 3D (x, y, and z coordinates). Although a surface profile is essentially 2D, the parametric formulation provides a solution for both 3D lines and planes. The solution to the LS best-fit line in its parametric form using the Lagrange multiplier method is described in Forbes (1989), and our discussion here is merely an elaboration of his presentation with illustrative examples.

Therefore we will now consider the line in its parametric form,

$$\frac{(x - x_o)}{a} = \frac{(z - z_o)}{b}, \tag{15.8}$$

with the unknowns being a point on the line (x_o, z_o) and the direction cosines (a, b), where $a^2 + b^2 = 1$. The solution derived for this problem is easily scalable to the third dimension.

The normal distance from any point $B\ (x(i), z(i))$ to the line is given by (see Fig. 15.1):

$$\begin{aligned} d(i) &= \sqrt{AB^2 - AC^2} \\ AB &= \sqrt{(x(i)-x_o)^2 + (z(i)-z_o)^2} \\ AC &= a(x(i)-x_o) + b(z(i)-z_o) \end{aligned} \quad . \tag{15.9}$$

The objective function E is given by

$$E = \sum_{i=1}^{n} d(i)^2 = \sum_{i=1}^{n} (x(i)-x_o)^2 + (z(i)-z_o)^2 - (a(x(i)-x_o) + b(z(i)-z_o))^2 \, . \tag{15.10}$$

Therefore, the problem of determining the parameters of the LS line can be stated as: minimize $E = \sum_{i=1}^{n} d(i)^2$ to determine the unknowns $(x_o,\ z_o)$ and $(a,\ b)$, subject to the equality constraint $G = a^2 + b^2 - 1 = 0$.

First, we observe that the deviation function $d(i)$ is not linear in the unknowns. Second, the solution for $(a,\ b)$ has to satisfy the constraint $G = 0$. Such problems are referred to as constrained optimization problems. To solve this, we will have to borrow a few tricks from advanced mathematics. We illustrate the procedure in the following sequence of steps:

Step 1: centroid as a point on the line. As seen from Sect. 15.4, the centroid of the data is a point on the LS line. We will translate the original data so that the centroid is now the origin of our coordinate system. The advantage in doing this is a simplification in our subsequent mathematics because $(x_o,\ z_o)$ is now $(0,0)$.

Step 2: solving the constrained minimization problem. Because the origin is a point on the line, the problem has been simplified to determining $(a,\ b)$ by minimizing

$$E = \sum_{i=1}^{n} x(i)^2 + z(i)^2 - (ax(i) + bz(i))^2 \, , \tag{15.11}$$

subject to the equality constraint $G = a^2 + b^2 - 1 = 0$. Unlike the slope-intercept form for the line, we cannot determine a solution for $(a,\ b)$ by setting $\nabla E = 0$ because of the constraint $G = 0$. One approach to solving this problem would be to set $b = \sqrt{1-a^2}$ (or for a if the line is nearly horizontal), and solve for a by setting $\nabla E = 0$. This approach clearly becomes cumbersome for a line in 3D. The alternate recommended approach involves the method of Lagrange multiplier. We explain the method in the Sect. 15.6. A robust treatment of constrained optimization such as

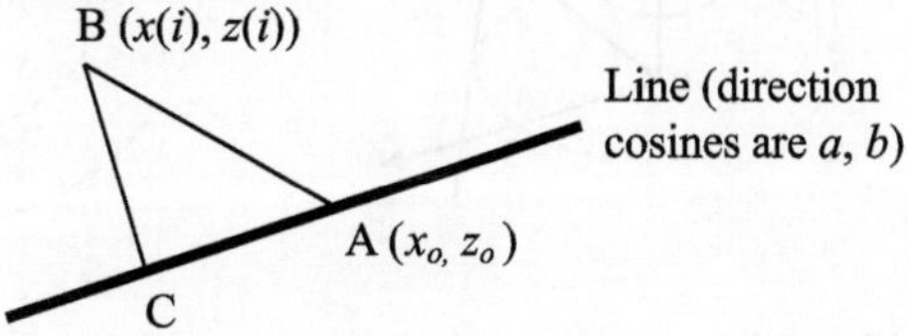

Fig. 15.1 Distance of a point to a line

the Lagrange multiplier method can be found in standard optimization texts such as (Reklaitis *et al.*, 1983; Rao 1996).

15.6 Lagrange Multiplier Method

Consider the constrained minimization problem $E = u^2 + v^2$, such that $u + v = 3\sqrt{2}$. If there are no constraints, the minimum value for E occurs when $u = 0$ and $v = 0$. This is easily seen because E is a sum of squared terms and therefore cannot take any negative values. In the presence of the constraint, $u = 0$, $v = 0$ is not a valid solution. Therefore, the problem is to determine the minima subject to the constraint. In Fig. 15.2, the objective function E is plotted as a function of u and v. The constraint is also plotted in the $u - v$ plane. We first attempt a graphical solution. Let us start by finding out (u, v) assuming $E = 25$. The line $u + v = 3\sqrt{2}$ intersects the circle $E = u^2 + v^2 = 25$ at two points and therefore a feasible solution does exist for (u, v). However, it is not the optimal solution as is evident from the figure. Only at the point where the constraint line is tangent to the circle $E = u^2 + v^2$ do we have an optimal solution. This occurs when $E = 9$. The constraint line does not intersect any smaller circle and so no feasible solutions are possible when $E < 9$.

From the Fig. 15.2, we notice the following important property at the solution point: "the gradient of the constraint line and the gradient of the surface are along the same direction." In other words, the constraint is a tangent to the surface. The gradient of the surface is given by $(2u, 2v)$ and the gradient of the constraint is $(1, 1)$. Therefore, when they are both along the same direction, we have $(2u, 2v) = \lambda(1, 1)$. The factor λ is introduced because the gradients only have to be along the same direction, not necessarily of the same magnitude. The factor λ is called a Lagrange multiplier. Therefore, the solution is $u = \lambda/2$, $v = \lambda/2$. And for this problem, the only feasible value for λ that satisfies the constraint is $\lambda = 3\sqrt{2}$.

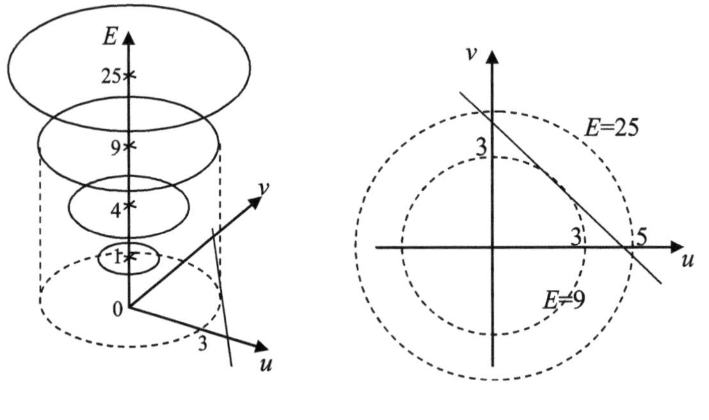

Fig. 15.2 Illustration of the Lagrange multiplier method

In summary, to minimize an objective function E, given an equality constraint G, we attempt to solve $\nabla E = \lambda \nabla G$. This approach is called the Lagrange multiplier method and we will use it to solve the minimization problem of the best-fit line, as shown next.

15.7 Back to the LS Line

Returning to the line problem, we apply the Lagrange multiplier method as follows.

1. First, we realize that the objective function given in Eq. 15.11 is actually the sum of two parts $E = E_1 - E_2$, where

$$E_1 = \sum_{i=1}^{n} x(i)^2 + z(i)^2 \tag{15.12}$$

and

$$E_2 = \sum_{i=1}^{n} (ax(i) + bz(i))^2 . \tag{15.13}$$

E_1 is a constant, while E_2 alone is a function of the unknown parameters. Therefore, instead of minimizing E, we maximize E_2 subject to the constraint $G = 0$.

2. According to the Lagrange multiplier method, we solve the system of equations $\nabla E_2 = \lambda \nabla G$, where

$$\nabla E_2 = \begin{bmatrix} \partial E_2/\partial a \\ \partial E_2/\partial b \end{bmatrix} \quad \text{and} \quad \nabla G = \begin{bmatrix} \partial G/\partial a \\ \partial G/\partial b \end{bmatrix} . \tag{15.14}$$

Evaluating the partial differentials of E and G, we have an eigen value problem $AV = \lambda V$, where

$$A = \begin{bmatrix} \sum_{i=1}^{n} x(i)^2 & \sum_{i=1}^{n} x(i)z(i) \\ \sum_{i=1}^{n} x(i)z(i) & \sum_{i=1}^{n} z(i)^2 \end{bmatrix} , \quad V = \begin{bmatrix} a \\ b \end{bmatrix} . \tag{15.15}$$

Notice that A can be rewritten in the form $M^T M$, where $M = \begin{bmatrix} x(1) & z(1) \\ \dots & \dots \\ x(n) & z(n) \end{bmatrix}$.

Therefore, the eigen value problem can be restated as $M^T MV = \lambda V$.

3. This eigen value problem has two solutions for λ, with each having a corresponding eigen vector V. To determine which eigen value to choose, consider the meaning of λ. Multiply the first row of $M^T MV = \lambda V$ by a, the second row by b and add the rows to obtain

$$\lambda = \left(a^2 \sum_{i=1}^{n} x(i)^2 + b^2 \sum_{i=1}^{n} z(i)^2 + 2ab \sum_{i=1}^{n} x(i)z(i) \right) = \sum_{i=1}^{n} (ax(i) + bz(i))^2 . \tag{15.16}$$

Thus, λ is the value of the objective function of E_2. In order to maximize E_2, we choose the larger of the two eigen values and the corresponding eigen vector as the direction cosines of the line.

4. And finally, we note that the singular values of M are identical to the eigen values of M^TM and therefore, we do not have to evaluate M^TM. The singular vector corresponding to the *largest* singular value of M represents the direction cosines of the line.
5. For lines in 3D, M is given by $M = \begin{bmatrix} x(1) & y(1) & z(1) \\ \dots & \dots & \dots \\ x(n) & y(n) & z(n) \end{bmatrix}$. Again, the singular vector corresponding to the largest singular value of M represents the direction cosines.

Example 15.4. Write a MATLAB function to compute the parameters of a LS best-fit line in 3D. The input argument is the data matrix of size $n \times 3$.

```
function [a,b,c] = line3D(M)
% M is a matrix of coordinates and is of size n x 3
n=size(M,1);
% find centroid
xc = sum(M(:,1))/n;
yc = sum(M(:,2))/n;
zc = sum(M(:,3))/n;
% build matrix A
for i=1:n
    A(i,1) = M(i,1)-xc;
    A(i,2) = M(i,2)-yc;
    A(i,3) = M(i,3)-zc;
end
% find eigen values and eigen vector corresponding
% to largest eigen value
[v,d] = eig(A'*A);
a1=[d(1,1);d(2,2);d(3,3)];
[y,i]= max(a1);
a = v(1,i); b = v(2,i); c = v(3,i);
```

15.8 LS Best-Fit Plane

The distance of any point $[x(i), y(i)]$ from the LS best-fit plane is given by

$$d(i) = ax(i) + by(i) + cz(i) , \tag{15.17}$$

where the direction cosines of the normal to the plane are (a, b, c). The objective function to be minimized is therefore

$$E = \sum_{i=1}^{n} d(i)^2 = \sum_{i=1}^{n} (ax(i) + by(i) + cz(i))^2 . \tag{15.18}$$

Recall that in the case of the line, we maximized the same objective function (E_2). Here we minimize the objective function E and therefore choose the singular vector corresponding to the smallest singular value of M.

15.9 Summary

We have discussed methods to determine parameters of LS best-fit lines in 2D and 3D, and best-fit planes. At the outset, determining parameters of a line would appear to be a fairly trivial problem. That is true for lines in a plane, but not so for lines in 3D or for planes. We borrowed some techniques from advanced mathematics, namely the method of Lagrange multiplier, to solve the optimization problem. Such best-fit line and plane fit problems appear during preprocessing of surface texture data due to measurement misalignment. In Chap. 16, we will discuss LS best-fit circles and spheres, also applicable during preprocessing of surface texture data from cylindrical and spherical surfaces.

Exercises

15.1. Generate a sinusoidal profile z defined as $z = \sin(\frac{2\pi x}{2}) + \frac{x}{8}, 0 \leqslant x < 8$ mm, (use 0.01 mm interval along x). Evaluate and remove the best-fit line. Why is the slope of the line not equal to $1/8$?

15.2. Generate a 3D surface as follows: $z = \sin(\frac{2\pi x}{2}) + \sin(\frac{2\pi y}{2}) + \frac{x+y}{8}$, $0 \leqslant x < 8$ mm, $0 \leqslant y < 8$ mm (use 0.1 mm interval in x and y). Evaluate and remove the best-fit plane. What is slope of the plane along x and y?

15.3. Verify that the centroid of a set of data is indeed a point on the LS line (Sect. 15.4).

15.4. In Example 15.3, explicitly evaluate matrices $A^T AP$ and $A^T B$, and solve the resulting set of two simultaneous equations in two variables in MATLAB. (Hint: Find out how to solve simultaneous equations in MATLAB. Alternately, derive the expressions for the unknowns as in Eq. 15.4 and then use MATLAB to determine the values for the unknowns.)

15.5. Write a MATLAB function to compute the parameters of a best-fit plane. The argument to the function is a matrix of data (size $n \times 3$). The output arguments are the centroid and direction cosines of the normal.

References

Forbes, A.B. 1989, *Least-Squares Best-Fit Geometric Elements*, NPL Report DITC 140/89, National Physical Laboratory. Teddington, UK.

Heath, M. 2002, *Scientific Computing: an Introductory Survey*, 2nd edn, McGraw-Hill. New-York, NY, USA.

Kreyszig, E. 2005, *Advanced Engineering Mathematics*, 9th ednWiley. New York, NY, USA.

Rao, S. 1996, *Engineering Optimization: Theory and Practice*, 3rd edn, Wiley-Interscience. New York, NY, USA.

Reklaitis, G.V., Ravindran, A. and Ragsdell, K.M. 1983, *Engineering Optimization: Methods and Applications*, 1st edn, Wiley-Interscience. New York, NY, USA.

Stoer, J., Bulirsch, R., Bartels, R. (translator), Gautschi, W. (translator) and Witzgall, C. (translator) 2002, *Introduction to Numerical Analysis*, 3rd edn, Springer. New York, NY, USA.

Chapter 16
Non-linear Least-Squares I: Introduction

16.1 Introduction

The analysis of surface texture measurements made on curved surfaces involves a preprocessing step where the nominal geometry is often removed prior to filtering. Circles, spheres, and cylinders are some of the most commonly found geometries. We therefore discuss methods for obtaining these LS best-fit geometries in this chapter.

In determining best-fit parameters for lines and planes, we were able to exploit the fact that the objective function upon partial differentiation produces linear equations in the unknowns. In many situations, such as for circles, spheres, and cylinders, this is not true and we are therefore forced to find solutions by either approximating the objective function and/or using iterative techniques. In this chapter, we discuss some iterative algorithms that are used in the solution process for determining LS best-fit circles, spheres, and cylinders. We use circles in a plane as an example to illustrate the methods.

16.2 Formulating the Circle in a Plane Problem

A circle can be specified by its center (x_c, z_c) and its radius r. We use variables (x, z) instead of (x, y) because the measurement data to which we fit a non-linear LS circle is anticipated to be a profile in the x–z plane, with measured surface profile height deviations along z. A LS best-fit circle is also fit to roundness profiles in the x–y plane, but the data is generally collected in polar coordinates and linear approximations are used as seen in Chap. 18.

Any point $(x(i), z(i))$ on the circle satisfies

$$(x(i) - x_c)^2 + (z(i) - z_c)^2 = r^2. \tag{16.1}$$

The radial distance (deviation from circle) from any point $(x(i), z(i))$ to a circle specified by x_c, z_c, and r is given by

$$d(i) = R(i) - r\,, \tag{16.2}$$

where

$$R(i) = \sqrt{(x(i) - x_c)^2 + (z(i) - z_c)^2}\,. \tag{16.3}$$

The objective function E is therefore given by

$$E = \sum_{i=1}^{n} d(i)^2\,. \tag{16.4}$$

Partial differential of E with respect to x_c, z_c, and r results in three non-linear equations in the unknowns that cannot be solved easily. In order to determine estimates of the parameters, we have to use iterative techniques such as those described next. We will require initial estimates of the unknown parameters, which will then be gradually refined. For now, we assume that initial estimates are available and proceed with a discussion of the iterative methods. In Chap. 17, we discuss a method for determination of initial estimates, also due to Forbes (1989).

The methods discussed in subsequent sections are unconstrained optimization techniques. The Lagrange multiplier technique discussed in Chap. 15 was an example of a constrained optimization technique. Any standard reference text on optimization covers these topics in sufficient detail; suggested references include (Reklaitis *et al.*, 1983; Rao 1996).

16.3 The Steepest Descent Algorithm

The steepest descent algorithm is the simplest of all gradient search methods, and by itself is rarely used for these types of problems because of its slow convergence. We, however, discuss it here to demonstrate how gradient-search-based algorithms generally function, and also because of a very important "provably convergent" property. This implies that if there is a local minimum, the algorithm will detect it eventually, even if it takes large number of iterations. The key idea here is as follows: given an objective function to be minimized along with initial estimates for the unknown parameters, the estimates for the next iteration are determined by going along the direction of steepest descent, which is the negative gradient of the objective function at the current solution. To see why this is so, consider Example 16.1.

Example 16.1. Determine parameters u and v by minimizing $E = u^2 + v^2$.

Let the initial estimates for (u, v) be (α, β). In Fig. 16.1, each circle represents a curve of equal objective function value. Thus, $E = 0$ is a point on the origin, because the only possible values for u and v are $u = 0$ and $v = 0$. $E = 1$ is a circle of unit radius, and so on. Obviously, E cannot take negative values. Thus, given any

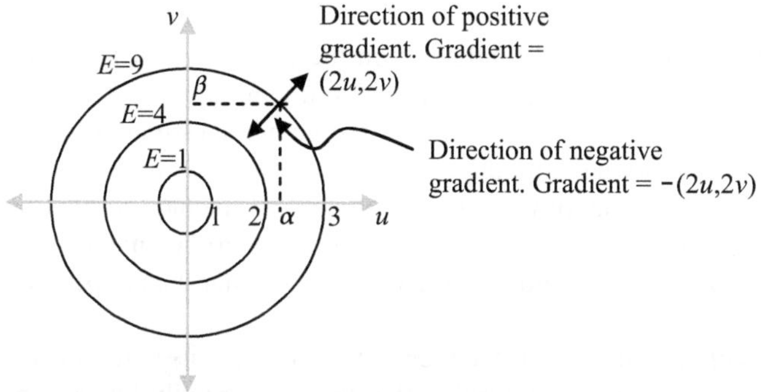

Fig. 16.1 Steepest descent algorithm illustration

initial estimate (α, β), the direction in which the function decreases most rapidly is along the negative gradient. For example, from (α, β), if we travel along the negative gradient by unit magnitude, the function value is 4. However along any other direction, the function will have values between 4 and 16. Traveling horizontally to the right by unit magnitude, the function is larger than 9. Traveling one unit to the left horizontally, the value of the function is between 4 and 9, but never as small as 4. The question then is: how far do we travel in this direction of negative gradient? In fact, in this example, if we travel far enough, we will directly strike the origin, which happens to be the minimum, but in other situations, more iterations will be necessary. Therefore we will choose an arbitrary factor that we later tune during our iterations to determine how far to travel along the negative gradient.

We will now illustrate the method of steepest descent for the circle problem. The objective function E is defined in Eq. 16.4. Assume that initial estimates for the center (x_c, z_c) and its radius r are (x_{co}, z_{co}) and r_o. We determine the gradient of the objective function E at the initial estimate, which is given by

$$\nabla E = \begin{bmatrix} -2\sum\limits_{i=1}^{n}(R_o(i) - r_o)(x(i) - x_{co})/R_o(i) \\ -2\sum\limits_{i=1}^{n}(R_o(i) - r_o)(z(i) - z_{co})/R_o(i) \\ -2\sum\limits_{i=1}^{n}(R_o(i) - r_o)\,. \end{bmatrix} \tag{16.5}$$

where $R_o(i)$ is the value of $R(i)$ evaluated at the current estimates. Then, the parameters at the end of the iteration are given by

$$\begin{bmatrix} x_{c1} \\ z_{c1} \\ r_1 \end{bmatrix} = \begin{bmatrix} x_{co} \\ z_{co} \\ r_o \end{bmatrix} - \lambda \begin{bmatrix} 2\sum\limits_{i=1}^{n}(R_o(i) - r_o)(x(i) - x_{co})/R_o(i) \\ 2\sum\limits_{i=1}^{n}(R_o(i) - r_o)(z(i) - z_{co})/R_o(i) \\ 2\sum\limits_{i=1}^{n}(R_o(i) - r_o) \end{bmatrix}, \tag{16.6}$$

where λ is some arbitrary scale factor parameter. It is set at a reasonably small value so as not to miss the minimum completely; on the other hand, it is not set too small so that it takes too many iterations to arrive at the minimum. After each iteration, the preceding values of parameters are reset as the initial estimates for the next iteration. This process continues until the convergence criteria are met. Convergence can be defined by any number of ways. Common methods include computing E and specifying changes in E between consecutive iterations to be smaller than a specified threshold, or the absolute values of E itself to be smaller than a specified threshold, *etc.*

The primary issue with this method, as we mentioned earlier, is its slow convergence. We will illustrate next another algorithm that has substantially better convergence, especially if the initial estimates are reasonably accurate. This method is called the Gauss–Newton method, and the important difference between the two is the direction in which to search for the solution. While the steepest descent uses the direction of negative gradient, the Gauss–Newton searches in the direction of the tangent to the surface.

16.4 The Gauss–Newton Algorithm

The application of the Gauss–Newton algorithm for non-linear LS fitting in metrology is due to Forbes of the National Physical Laboratory, UK; the discussion and equations below are an elaboration from his report (Forbes 1989).

The central idea in this iterative procedure is that, given an initial set of estimates of the unknown parameters (center and radius for a circle), we can linearize the deviation function in the neighborhood of the initial estimate and find new estimates that will minimize the linear objective function. We then use these estimates as the initial estimate for the next iteration. This is illustrated in Fig. 16.2 where we start with an initial estimate $x = x_o$ to locate the root of the function $f(x) = 0$. At $x = x_o$, we linearize the function by its tangent [Taylor series expansion ignoring higher

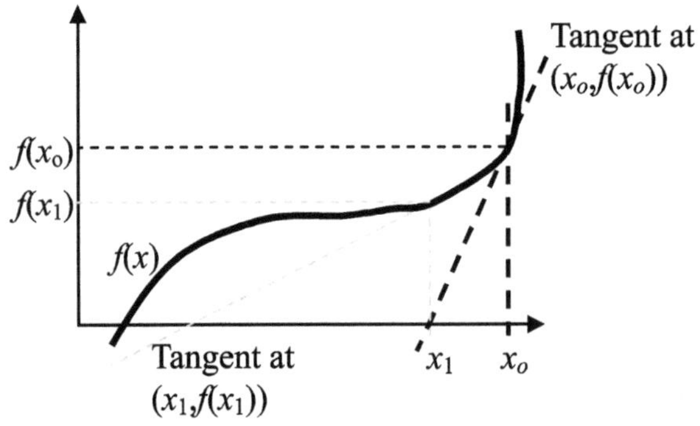

Fig. 16.2 The Gauss–Newton algorithm

order terms, $f'(x_0) \neq 0$], so that

$$\begin{aligned} f(x) &= f(x_o) + (x - x_o) f'(x_o) + \frac{(x - x_o)^2}{2} f''(x_o) + \dots \\ &\approx f(x_o) + (x - x_o) f'(x_o) \,. \end{aligned} \tag{16.7}$$

The estimate x_1 for the next iteration is obtained by finding the root of this new linear function. Thus, by solving $f(x) = 0$, we get

$$x_1 = x_o + f(x_o)/f'(x_o) \,. \tag{16.8}$$

In the case of the circle-fitting problem, the deviation of the ith data point is given by

$$d(i) = R(i) - r \,, \quad R(i) = \sqrt{(x(i) - x_c)^2 + (z(i) - z_c)^2} \,. \tag{16.9}$$

Given an initial estimate for the center and radius, (x_{co}, z_{co}) and r_o, we linearize the function using Taylor's expansion as

$$d(i) \approx R_o(i) - r_o + (x_{c1} - x_{co}) \frac{\partial d(i)}{\partial x_{co}} + (z_{c1} - z_{co}) \frac{\partial d(i)}{\partial z_{co}} + (r_1 - r_o) \frac{\partial d(i)}{\partial r_o} \tag{16.10}$$

where (x_{c1}, z_{c1}) and r_1 are the new estimates at the end of the iteration, and

$$R_o(i) = \sqrt{(x(i) - x_{co})^2 + (z(i) - z_{co})^2} \tag{16.11}$$

is the value of $R(i)$ evaluated at the initial estimates.

Having linearized the objective function, we realize that we have n equations and three unknowns, a simple linear LS problem. Recall from Chap. 15 that Eq. 16.10 can be expressed in matrix form as $JP = -D$, and that its LS solution is obtained by solving the normal form $J^T JP = -J^T D$. J, P, and D are given by

$$J = \begin{bmatrix} \frac{\partial d(1)}{\partial x_{co}} & \frac{\partial d(1)}{\partial z_{co}} & \frac{\partial d(1)}{\partial r_o} \\ \dots & \dots & \dots \\ \frac{\partial d(n)}{\partial x_{co}} & \frac{\partial d(n)}{\partial z_{co}} & \frac{\partial d(n)}{\partial r_o} \end{bmatrix} , \quad P = \begin{bmatrix} x_{c1} - x_{co} \\ z_{c1} - z_{co} \\ r_1 - r_o \end{bmatrix} , \quad D = \begin{bmatrix} R_o(1) - r_o \\ \dots \\ R_o(n) - r_o \end{bmatrix} . \tag{16.12}$$

Matrix J is called the Jacobian matrix; it contains the partial derivatives of the distance function with respect to the unknowns evaluated at the current estimates for the center and radius. The partial derivatives are given by

$$\begin{aligned} \frac{\partial d(i)}{\partial x_{co}} &= -(x(i) - x_{co})/R_o(i) \\ \frac{\partial d(i)}{\partial z_{co}} &= -(z(i) - z_{co})/R_o(i) \\ \frac{\partial d(i)}{\partial r_o} &= -1 \,. \end{aligned} \tag{16.13}$$

The iterations can be repeated until either (a) a maximum number limit for iterations has been reached, (b) the objective function E does not change appreciably from iteration to iteration, (c) the unknown parameters do not change appreciably, or (d) the absolute value of the objective E has fallen below a threshold.

Example 16.2. Write a MATLAB function that evaluates the parameters of a circle using the Gauss–Newton Method.

```
function [xc,zc,r] = LSCircleInPlaneGN(M,xco,zco,ro)
% M is a matrix of data, of size nx2; xco, zco and ro are
% initial estimates
eps = 1e-8;
flag = 0;
n = size(M,1);
Eo = sum((sqrt((M(:,1)-xco).^2+(M(:,2)-zco).^2)-ro).^2);
% Gauss-Newton iteration for LS best fit circle
t = 1;
while flag == 0
    % Jacobian
    for i=1:n
       R = sqrt((M(i,1)-xco)^2+(M(i,2)-zco)^2);
       J(i,1) = -(M(i,1)-xco)/R;J(i,2)=-(M(i,2)-zco)/R;J(i,3)=-1;
       D(i,1) = -(R-ro);
    end
    P = J\D;
    xco = xco + P(1); zco = zco + P(2); ro = ro + P(3);
    E1 = sum((sqrt((M(:,1)-xco).^2+(M(:,2)-zco).^2)-ro).^2);
    if abs(Eo-E1) < eps
        flag = 1;
    end
    Eo = E1;
    disp(['iteration # ' num2str([t xco zco ro Eo])]);
    t = t + 1;
end
xc = xco; zc = zco; r = ro;
```

16.5 The Levenberg–Marquardt Algorithm

We next consider a third algorithm that is a combination of the steepest descent and the Gauss–Newton. Called the Levenberg–Marquardt algorithm, it uses the same linearized objective function used in the Gauss–Newton, Eq. 16.10.

The difference from the Gauss–Newton algorithm lies in that instead of finding a solution using $J^TJP = -J^TD$, where the Jacobian is as described in Sect. 16.4, we attempt to find a solution using a modified equation, $(J^TJ + \lambda R^TR)P = -J^TD$, where λ is the Levenberg–Marquardt parameter and R^TR is the identity matrix plus the diagonal elements of J^TJ. The motivation for this additional matrix is as follows: when λ is zero, the method is identical to the Gauss–Newton algorithm, and the solution therefore follows the direction of the tangent. However, as λ is made larger, the diagonal elements of the Jacobian now dominate in comparison to the off-diagonal terms, which implies that the solution is searched along the negative gradient, making it similar to the steepest descent. Thus, by varying λ during the iterations, we can control the direction in which to search for the solution. The Levenberg–Marquardt algorithm for metrology applications is described by Shakarji (1998). Below is an adaptation of Shakarji's algorithm[1].

The algorithm is as follows:

1. Determine initial estimates for the center and radius.
2. Evaluate the objective function E at the initial estimates.
3. Set λ to some very small value, say 0.0001.
4. Build matrices J and D as before.
5. Start a second loop to decide whether to increment or decrement λ.
   a. Evaluate $J^T J + \lambda R^T R$.
   b. Solve $(J^T J + \lambda R^T R)P = -J^T D$ to determine new estimate of the center and radius.
   c. Evaluate the objective function E at the new set of parameter estimates.
   d. If objective function value is smaller than before,
      i. $\lambda \Rightarrow \lambda \times 0.04$.
      ii. Set the current estimates of the parameters as the initial estimate for the next iteration.
      iii. Terminate this loop and go back to step 4.
   e. Else
      iv. $\lambda \Rightarrow \lambda \times 10$.
      v. Do not accept the current estimates as valid.
      vi. Go back to step (a) in this loop with the old values for initial estimates, but with the new value for λ.

If the difference between the objective function values between any two consecutive iterations is smaller than a stated threshold, the algorithm may be terminated.

16.6 Summary

In this chapter, we have focused primarily on solution methodology for non-linear LS best-fit problems. We have discussed three iterative techniques using the simple geometry of a circle in a plane. The ideas discussed here can easily be extended to another dimension for spheres and cylinders, as we will see in Chap. 17. The solution strategy involves iterative procedures with a reasonably good initial estimate. The initial estimate can be obtained either mathematically or by an educated guess. Generally, the steepest descent algorithm guarantees convergence, but it is

[1] Official contribution of the National Institute of Standards and Technology; not subject to copyright in the United States.

not usually the preferred choice because of its poor speed. The Gauss–Newton and the Levenberg–Marquardt algorithm are generally the more preferred approaches. MATLAB provides a function `lsqnonlin` that may be used to implement these algorithms. We highlight the use of this function through an exercise (see Exercise 16.6 and its solution in the answer key).

Exercises

16.1. Write a program to determine the estimates for a LS circle using the steepest descent technique.

16.2. Write a program to determine the estimates for a LS circle using the Levenberg–Marquardt technique.

16.3. Determine the parameters of the LS best-fit circle using each of the three algorithms for the matrix A containing six (x,z) coordinate pairs.

x	10.1852	5.3514	−4.7267	−9.7776	−4.6527	5.3107
z	0.3974	9.1387	8.9215	0.4401	−8.5738	−8.1704

Plot the data and determine estimates for the center and radius. Use $(0,0)$ and 10 as initial estimates for the center and radius.

16.4. In the previous exercise, does the steepest descent algorithm converge for $\lambda = 2$? What is the advantage of increasing λ? Does the algorithm converge when $\lambda = 0.01$, and if so, in how many iterations? What is the advantage of decreasing λ?

16.5. Intentionally perturb the initial estimates by small amounts and then recompute the center and radius using each of the three algorithms. Which technique takes the most time to execute?

16.6. Determine the parameters of the LS best-fit circle in Exercise 16.3 using MATLAB's `lsqnonlin` function.

References

Forbes, A.B. 1989, *Least-Squares Best-Fit Geometric Elements*, NPL Report DITC 140/89, National Physical Laboratory. Teddington, UK.

Rao, S. 1996, *Engineering Optimization: Theory and Practice*, 3rd edn, Wiley-Interscience. New York, NY, USA.

Reklaitis, G.V., Ravindran, A. and Ragsdell, K.M. 1983, *Engineering Optimization: Methods and Applications*, 1st edn, Wiley-Interscience. New York, NY, USA.

Shakarji, C. 1998, 'Least-squares fitting algorithms of the NIST Algorithm Testing System', *Journal of Research of the NIST*, vol. 103, no. 6, pp. 633–641.

Chapter 17
Non-linear Least-Squares II: Circle, Sphere, and Cylinder

17.1 Introduction

In Chap. 16, we discussed three iterative techniques for solving non-linear LS problems. We illustrated each of the methods using the case of the circle-fitting problem. In this chapter, we extend that discussion to spheres and cylinders, but first, we discuss a technique for evaluating initial estimates for the circle.

17.2 Initial Estimates for Center and Radius of Circle

Forbes (1989) presents a method to determine initial estimates for the center and radius by simplifying the objective function such that the resulting partial differentials are linear in the unknowns. The strategy is to approximate E as shown:

$$E = \sum_{i=1}^{n} d(i)^2 = \sum_{i=1}^{n} (R(i) - r)^2 \approx \sum_{i=1}^{n} (R(i)^2 - r^2) . \tag{17.1}$$

Substituting for $R(i)$ and expanding, we have the following linear system of equations in x_c, y_c, and ρ as shown below:

$$\begin{aligned} E &= \sum_{i=1}^{n} ((x(i) - x_c)^2 + (z(i) - z_c)^2 - r^2) \\ &= \sum_{i=1}^{n} (-2x(i)x_c - 2z(i)z_c + \rho + x(i)^2 + z(i)^2) = 0 , \end{aligned} \tag{17.2}$$

where $\rho = x_c^2 + z_c^2 - r^2$.

E can now be re-written in matrix form $AP = B$, where

$$A = \begin{bmatrix} 2x(1) & 2z(1) & -1 \\ \dots & \dots & \dots \\ 2x(n) & 2z(n) & -1 \end{bmatrix} , \quad P = \begin{bmatrix} x_c \\ z_c \\ \rho \end{bmatrix} , \quad B = \begin{bmatrix} x(1)^2 + z(1)^2 \\ \dots \\ x(n)^2 + z(n)^2 \end{bmatrix} . \tag{17.3}$$

Note that E is a sum of squared terms and therefore each individual term $(d(i)^2)$ cannot be smaller than zero. The matrix form in Eq. 17.3 is obtained by equating each term within the parentheses in Eq. 17.2 to zero and solving the unknowns. An estimate of r is then obtained from the estimate for ρ.

Example 17.1. Simulate a surface profile 2 mm long, acquired on the surface of a circle of radius 25 mm, containing only random noise. Assume spacing is 1 μm.

The surface profile is acquired along, and normal to, the chord of the circle; that is, x values are measured along the chord, while the deviations are measured normal to the chord. From Fig. 17.1, we can determine z coordinates of points on the circle as a function of x if the origin is O. However, for the stylus instrument dragging a profile from A to B, the origin is A.

From Fig. 17.1, we see that

$$AC = l/2$$

$$CO = \sqrt{r^2 - (l/2)^2} .$$

Because the coordinate system has to be shifted to point A, we can write the equation for the circle as

$$(x - AC)^2 + (z + CO)^2 = r^2 ,$$

from which we get

$$z = -CO + \sqrt{r^2 - (x - AC)^2} .$$

The following MATLAB code segment illustrates the process of generating z as a function of x along with some superimposed random noise scaled to maximum amplitude of 1 μm.

```
% Simulation of data on surface of circle
x = 0:0.1:2;
z = sqrt(25^2 - (x-1).^2) - sqrt(25^2 - 1^2)
    + rand(1,size(x,2))*0.001;
plot(x,z)
```

Example 17.2. Verify that the best-fit radius is close to the theoretical radius in Example 17.1

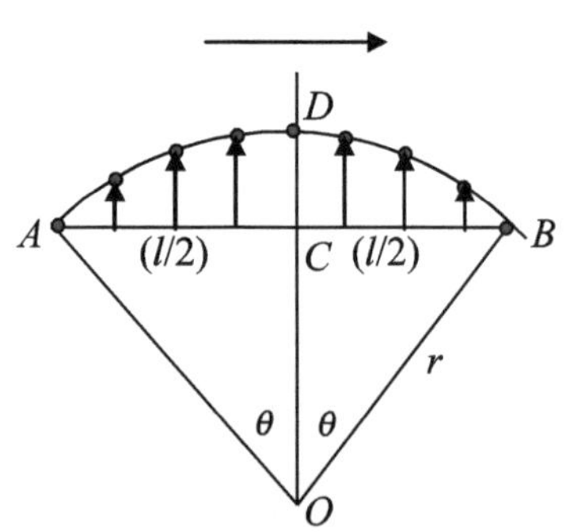

Fig. 17.1 Simulating a surface profile on the surface of a cylinder

Continuing from Example 17.1, we have

```
M = [x' z'];
[xc,zc,r] = LSCircleInPlaneGN(M,1,-24,4);
```

where $(1,-24)$ and 4 are provided as initial estimates for the center and radius. We leave it as an exercise to the reader to write a program that computes the initial estimates as described in Sect. 17.2.

Example 17.3. Assume that a perfectly circular thin disk is mounted on a roundness instrument with an eccentricity $(a,\ b)$. In Chap. 18, we will show that the radial deviations measured by the probe $R(\theta)$ can be approximated as $a\cos\theta + b\sin\theta + r$, where r is the part radius. For $(a,b) = (0.2,\ -0.5)$ µm, and $r = 7$ µm, generate and plot the roundness profile assuming a point is sampled every degree.

The roundness profile can be generated as shown below. It is plotted in Fig. 17.2.

```
% Generating a roundness profile
theta = (0:1:359)'*pi/180;
a = 0.2; b = -0.5; r = 7;
R = a.*cos(theta) + b.*sin(theta) + r;
polar(theta,R)
```

Example 17.4. Determine the best-fit parameters of the circle in Example 17.3, using any of the iterative methods discussed in Chap. 16.

We first note that the roundness data in Example 17.3 is in polar coordinates while the non-linear LS algorithms require inputs in Cartesian coordinates. We will therefore convert the polar data to Cartesian, and use the Gauss–Newton algorithm as shown below

```
x = R.*cos(theta);
y = R.*sin(theta);
[xc,yc,r] = LSCircleInPlaneGN([x y],0,0,2); % initial estimates
                                            % are (0,0) and 2.
```

Notice that the center, $(0.19971,\ -0.49928)$, is very close to the eccentricity, $(0.2,\ -0.5)$, as expected.

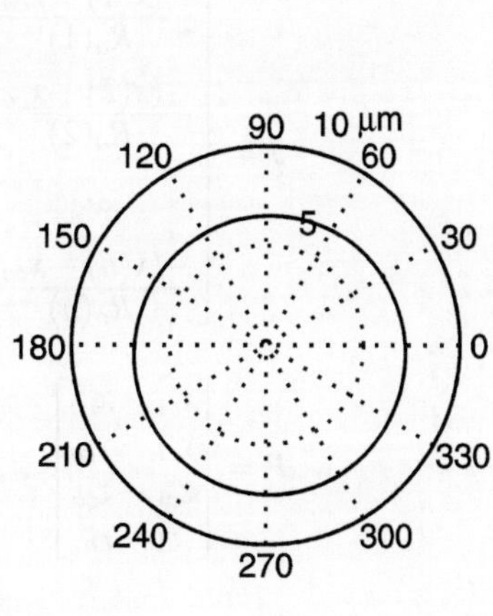

Fig. 17.2 Roundness profile in presence of eccentricity

17.3 Best-Fit Sphere

The sphere is a simple 3D extension to the circle-fitting problem, and we will therefore only discuss it briefly here. The problem is to minimize

$$E = \sum_{i=1}^{n} d(i)^2 = \sum_{i=1}^{n} (R(i) - r)^2 , \tag{17.4}$$

where

$$R(i) = \sqrt{(x(i) - x_c)^2 + (y(i) - y_c)^2 + (z(i) - z_c)^2} . \tag{17.5}$$

An initial solution can be found in a fashion identical to that for the circle by approximating the objective function as

$$E = \sum (R(i) - r)^2 \approx \sum (R(i)^2 - r^2) , \tag{17.6}$$

where $\rho = x_c^2 + y_c^2 + z_c^2 - r^2$. The initial estimates for center and radius are obtained by solving the system $AP = B$ (solve the corresponding normal form $A^T AP = A^T B$), where

$$A = \begin{bmatrix} 2x(1) & 2y(1) & 2z(1) & -1 \\ 2x(2) & 2y(2) & 2z(2) & -1 \\ \dots & \dots & \dots & \dots \\ 2x(n) & 2y(n) & 2z(n) & -1 \end{bmatrix}, \quad P = \begin{bmatrix} x_c \\ y_c \\ z_c \\ \rho \end{bmatrix} \text{ and } B = \begin{bmatrix} x(1)^2 + y(1)^2 + z(1)^2 \\ x(2)^2 + y(2)^2 + z(2)^2 \\ \dots \\ x(n)^2 + y(n)^2 + z(n)^2 \end{bmatrix} . \tag{17.7}$$

After determining the initial estimates, we can choose from any of the iterative methods discussed earlier to refine our initial estimates to the true LS solution. Assuming we use the Gauss–Newton or the Levenberg–Marquardt algorithms, we have to evaluate the Jacobian matrix J and solve the system $JP = -D$ or $(J^T J + \lambda R^T R)P = -J^T D$ as the case may be. For the sphere, J, P, and D are given by

$$J = \begin{bmatrix} \frac{-(x(1) - x_{co})}{R_o(1)} & \frac{-(y(1) - y_{co})}{R_o(1)} & \frac{-(z(1) - z_{co})}{R_o(1)} & -1 \\ \frac{-(x(2) - x_{co})}{R_o(2)} & \frac{-(y(2) - y_{co})}{R_o(2)} & \frac{-(z(2) - z_{co})}{R_o(2)} & -1 \\ \dots & \dots & \dots & \dots \\ \frac{-(x(n) - x_{co})}{R_o(n)} & \frac{-(y(n) - y_{co})}{R_o(n)} & \frac{-(z(n) - z_{co})}{R_o(n)} & -1 \end{bmatrix} \tag{17.8}$$

$$P = \begin{bmatrix} x_{c1} - x_{co} \\ y_{c1} - y_{co} \\ z_{c1} - z_{co} \\ r_1 - r_o \end{bmatrix}, \quad D = \begin{bmatrix} d(1) \\ \dots \\ d(n) \end{bmatrix} . \tag{17.9}$$

17.4 Best-Fit Cylinder

A cylinder can be specified by a point on the axis (x_c, y_c, z_c), direction cosines of the axis (a, b, c) and the radius r. We do not attempt a generalized procedure for the evaluation of initial estimates. Forbes does briefly mention a possible approach (Forbes 1989). The distance from any point $(x(i), y(i), z(i))$ to the axis of the cylinder can be given by

$$d(i) = R(i) - r\,, \tag{17.10}$$

where

$$R(i) = \frac{\sqrt{u(i)^2 + v(i)^2 + w(i)^2}}{\sqrt{a^2 + b^2 + c^2}} \tag{17.11}$$

and

$$\begin{aligned} u(i) &= c(y(i) - y_c) - b(z(i) - z_c) \\ v(i) &= a(z(i) - z_c) - c(x(i) - x_c)\;. \\ w(i) &= b(x(i) - x_c) - a(y(i) - y_c) \end{aligned} \tag{17.12}$$

The objective function is given by

$$E = \sum_{i=1}^{n} d(i)^2\,. \tag{17.13}$$

As before, we linearize the objective function to

$$\begin{aligned} d(i) =& R_o(i) - r_o + (x_{c1} - x_{co})\frac{\partial d(i)}{\partial x_{co}} + (y_{c1} - y_{co})\frac{\partial d(i)}{\partial y_{co}} + (z_{c1} - z_{co})\frac{\partial d(i)}{\partial z_{co}} \\ &+ (a_1 - a_o)\frac{\partial d(i)}{\partial a_o} + (b_1 - b_o)\frac{\partial d(i)}{\partial b_o} + (c_1 - c_o)\frac{\partial d(i)}{\partial c_o} + (r_1 - r_o)\frac{\partial d(i)}{\partial r_o}\,, \end{aligned} \tag{17.14}$$

and solve the resulting system of n equations in seven variables expressed in the form $JP = -D$ or $(J^T J + \lambda R^T R)P = -J^T D$, as the case may be. The matrices J, P, and D are given by

$$J = \begin{bmatrix} \frac{\partial d(1)}{\partial x_{co}} & \frac{\partial d(1)}{\partial y_{co}} & \frac{\partial d(1)}{\partial z_{co}} & \frac{\partial d(1)}{\partial a_o} & \frac{\partial d(1)}{\partial b_o} & \frac{\partial d(1)}{\partial c_o} & \frac{\partial d(1)}{\partial r_o} \\ \cdots & \cdots & \cdots & \cdots & \cdot\cdot & \cdots & \cdots \\ \frac{\partial d(n)}{\partial x_{co}} & \frac{\partial d(n)}{\partial y_{co}} & \frac{\partial d(n)}{\partial z_{co}} & \frac{\partial d(n)}{\partial a_o} & \frac{\partial d(n)}{\partial b_o} & \frac{\partial d(n)}{\partial c_o} & \frac{\partial d(n)}{\partial r_o} \end{bmatrix} \tag{17.15}$$

$$P = \begin{bmatrix} (x_{c1} - x_{co}) \\ (y_{c1} - y_{co}) \\ (z_{c1} - z_{co}) \\ (a_1 - a_o) \\ (b_1 - b_o) \\ (c_1 - c_o) \\ (r_1 - r_o) \end{bmatrix} \quad \text{and} \quad D = \begin{bmatrix} d(1) \\ \cdots \\ d(n) \end{bmatrix}\,. \tag{17.16}$$

The partial differentials for Eq. 17.15 can be evaluated from Eqs. 17.10, 17.11, and 17.12.

17.5 Summary

LS best-fit cylinders and spheres are sometimes used for preprocessing surface texture data prior to filtering. Because circles, spheres, and cylinders are some of the most commonly found geometries, we discussed methods for determining parameters of their LS best-fit geometry in this and Chap. 16. Traditional filters, such as the Gaussian filter, are influenced by the nominal shape of the surface. In the presence of large form, such as found during the measurement of curved surfaces, the waviness or form profiles are influenced by the nominal shape. In fact, some recent developments such as the spline filter and the second-order Gaussian regression filter were developed to overcome this problem. The partial differentials for the Jacobian matrix for different geometries are listed by Shakarji (1998).

Exercises

17.1. Write a MATLAB program that computes the initial estimates for circles.

17.2. Write a MATLAB program that computes the initial estimates for spheres.

17.3. Write a MATLAB program that computes the Gauss–Newton estimate for spheres.

17.4. Assume that a 3D area map is acquired on the surface of a sphere. It can be simulated as shown below:

```
t=1;
for i=1:21
    for j=1:21
        x(i,j) = (i-1)*0.1; % in units of millimeters
        y(i,j) = (j-1)*0.1; % in units of millimeters
        z(i,j) = sqrt(25^2 - (x(i,j)-1).^2 - (y(i,j)-1).^2)
                 - sqrt(25^2 - 1^2) + rand(1,1)*0.001;
        M(t,1) = x(i,j);
        M(t,2) = y(i,j);
        M(t,3) = z(i,j);
        t = t+1;
    end
end
```

Obtain estimates of the best-fit sphere center and radius.

References

Forbes, A.B. 1989, *Least-Squares Best-Fit Geometric Elements*, NPL Report DITC 140/89, National Physical Laboratory. Teddington, UK.

Shakarji, C. 1998, 'Least-squares fitting algorithms of the NIST Algorithm Testing System', *Journal of Research of the NIST*, vol. 103, no. 6, pp. 633–641.

Chapter 18
Fitting Radius-Suppressed Circle Data

18.1 Introduction

A rotating spindle roundness instrument is generally used to determine the extent of out-of-roundness in a nominally circular (or cylindrical) part. The instrument, as we have discussed before, is insensitive to the actual diameter of the part. If the part is perfectly centered with respect to the spindle, the resulting deviations directly provide a measure of the roundness of the part. If the part is not centered, as we saw in Example 17.3, we have to determine the roundness from the best-fit circle. We have also seen in Chaps. 16 and 17 that determining the parameters of a best-fit involves using iterative techniques, and is therefore a little cumbersome.

In seminal work, Chetwynd (1979) and Chetwynd and Phillipson (1980) show that a circle can be approximated by a geometry called the limaçon. This approximation is valid in the presence of small eccentricities, a situation quite commonly found in roundness measurement scenarios where the part is fairly well centered, but not precisely so. The advantage of using the limaçon approximation is that its equation is linear in the unknown parameters, making it a fairly trivial process to determine the center and radius.

In this chapter, we will first discuss the determination of the LS circle parameters using the limaçon approximation. Subsequently, we will discuss limaçon-based cylinders (limaçoids), and finally, attempt to characterize the errors in this approximation in the presence of large eccentricities.

18.2 The Limaçon Approximation

Consider a circular part of radius r, mounted eccentrically on a spindle with eccentricity (a,b). Let the measured radial deviations of the probe as a function of angle θ be $R(\theta)$. From Fig. 18.1 (shown as if the probe is moving around the part instead

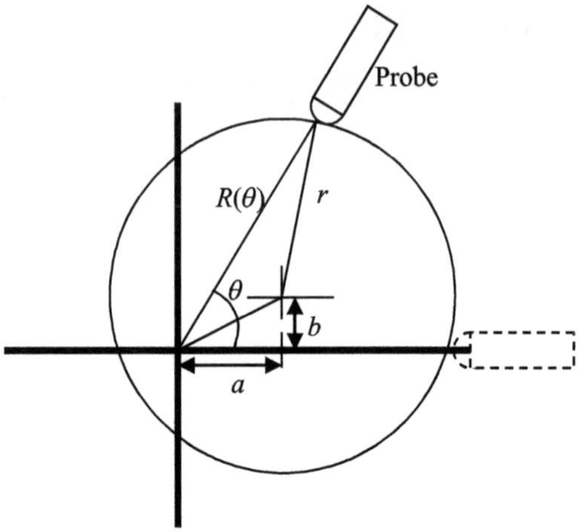

Fig. 18.1 The limaçon approximation. The part has an eccentricity given by (a,b). The probe reading $R(\theta)$ can be approximated as $R(\theta) = r + a\cos\theta + b\sin\theta$

of the part rotating), $R(\theta)$ can be given by

$$R(\theta) = a\cos\theta + b\sin\theta + \sqrt{r^2 - (a^2+b^2)\sin(\theta-\phi)}\,, \tag{18.1}$$

where

$$\phi = \tan^{-1}(b/a)\,. \tag{18.2}$$

Equation 18.1 can be expanded as

$$\begin{aligned} R(\theta) &= a\cos\theta + b\sin\theta + r\sqrt{1 - \frac{(a^2+b^2)\sin(\theta-\phi)}{r^2}} \\ &= a\cos\theta + b\sin\theta + r(1 - \frac{(a^2+b^2)\sin(\theta-\phi)}{2r^2} + \ldots)\,. \end{aligned} \tag{18.3}$$

Because a and b are small in comparison to r, we can ignore terms with higher powers of r in the denominator. Therefore we have,

$$R(\theta) = r + a\cos\theta + b\sin\theta\,. \tag{18.4}$$

18.3 LS Best-Fit Circle

Because Eq. 18.4 is linear in the unknowns (a,b,r), it is quite straightforward to find the parameters given a set of n radial deviations as a function of angular position. Equation 18.4 can be re-written in matrix form as $AP = B$, where

$$A = \begin{bmatrix} \cos\theta(1) & \sin\theta(1) & 1 \\ \ldots & \ldots & \ldots \\ \cos\theta(n) & \sin\theta(n) & 1 \end{bmatrix}, \quad P = \begin{bmatrix} a \\ b \\ r \end{bmatrix}, \quad B = \begin{bmatrix} R(1) \\ \ldots \\ R(n) \end{bmatrix}. \tag{18.5}$$

Note that although R is a function of θ [that functional relationship is expressed as $R(\theta)$], the array R that stores the radial distances (or deviations) is indexed from 1 to n, hence the notation $R(i)$. Each element $R(i)$ corresponds to the angle $\theta(i)$. We have seen in Chap. 15 that the LS solution is given by the form $A^T AP = A^T B$, which is

$$\begin{bmatrix} \sum_{i=1}^{n} \cos^2\theta(i) & \sum_{i=1}^{n} \cos\theta(i)\sin\theta(i) & \sum_{i=1}^{n} \cos\theta(i) \\ \sum_{i=1}^{n} \cos\theta(i)\sin\theta(i) & \sum_{i=1}^{n} \sin^2\theta(i) & \sum_{i=1}^{n} \sin\theta(i) \\ \sum_{i=1}^{n} \cos\theta(i) & \sum_{i=1}^{n} \sin\theta(i) & n \end{bmatrix} \begin{bmatrix} a \\ b \\ r \end{bmatrix} = \begin{bmatrix} \sum_{i=1}^{n} \cos\theta(i)R(i) \\ \sum_{i=1}^{n} \sin\theta(i)R(i) \\ \sum_{i=1}^{n} R(i) \end{bmatrix} \tag{18.6}$$

However, because data points are collected uniformly through out the circumference of the circle, we have

$$\begin{aligned} &\sum_{i=1}^{n} \cos\theta(i)\sin\theta(i) = 0 \\ &\sum_{i=1}^{n} \cos\theta(i) = 0 \\ &\sum_{i=1}^{n} \sin\theta(i) = 0 \\ &\sum_{i=1}^{n} \cos^2\theta(i) = n/2 \\ &\sum_{i=1}^{n} \sin^2\theta(i) = n/2 \end{aligned} \quad . \tag{18.7}$$

This gives the LS center and radius immediately as

$$\begin{aligned} a &= (2/n)\sum_{i=1}^{n} \cos\theta(i)R(i) \\ b &= (2/n)\sum_{i=1}^{n} \sin\theta(i)R(i) \\ r &= (1/n)\sum_{i=1}^{n} R(i) \end{aligned} \quad . \tag{18.8}$$

Example 18.1. Write a MATLAB program to compute the parameters of a limaçon. Assume that the n measured points span the entire circle at equally spaced intervals.

The following function computes the parameters of a best-fit limaçon and the radial out-of-roundness:

```
function [r, a, b, OOR] = lslimacon(R)
n = length(R); % R is the array of measured radial deviations
theta = (0:1:n-1)'*2*pi/n;
r = sum(R)/n;
a = 2*sum(R.*cos(theta))/n;
b = 2*sum(R.*sin(theta))/n;
R_ls = r + a*cos(theta)+b*sin(theta); % R_ls is the LS limacon
```

```
% plot the original data and the LS circle
polar(theta, R);
hold on
polar(theta, R_ls,'r-')
% Compute OOR
OOR = max(R-R_ls)-min(R-R_ls);          % out of roundness
[r a b OOR]
```

Example 18.2. Use the limaçon approximation and determine the best-fit parameters for the data in Example 17.3.

Continuing from Example 17.3, we can call the `lslimacon` function in Example 18.1, without converting the data to Cartesian:

```
[r,a,b,OOR] = lslimacon(R);
```

Notice that the out-of-roundness is zero, as expected, because the data generated was a perfect (no form) but eccentric circle.

18.4 LS Best-Fit Cylinder

Cylindrical form data is acquired by measuring circular traces at different heights. A mathematical model of cylindrical form was given by Kakino and Kitazawa (1978), and can be simplified here as shown. We can approximate the best-fit cylinder as a collection of limaçons that lie on an axis whose tilt is a function of z height. A limaçon describing a trace at any given height is given by

$$R(\theta) = r + a\cos\theta + b\sin\theta \,. \tag{18.9}$$

However, we also have to account for traces at other heights. Let z be the height, and e and f be the tilt parameters of the axis, and the limaçoid cylinder can then be represented by

$$R(\theta,z) = r + (a+ze)\cos\theta + (b+zf)\sin\theta \,. \tag{18.10}$$

Note that the functional relationship of R to θ and z is expressed as $R(\theta,z)$, and is not to be confused as a 2D matrix. In matrix form, let R, z, and θ be a vectors of length n whose first n_1 points represent the n_1 angular positions at $z=0$, the next n_1 points represent the angular positions at the next height, and so on for a total length of $n = n_1 \times n_2$ (n_2 traces). Then, we have

$$\begin{bmatrix} \cos\theta(1) & \sin\theta(1) & z(1)\cos\theta(1) & z(1)\sin\theta(1) & 1 \\ \cdots & \cdots & \cdots & \cdots & \cdots \\ \cos\theta(n) & \sin\theta(n) & z(n)\cos\theta(n) & z(n)\sin\theta(n) & 1 \end{bmatrix} \begin{bmatrix} a \\ b \\ e \\ f \\ r \end{bmatrix} = \begin{bmatrix} R(1) \\ \cdots \\ R(n) \end{bmatrix} . \tag{18.11}$$

This equation is of the form $AP = B$ and the LS solution for this equation can be obtained from its normal form, $A^T AP = A^T B$.

Example 18.3. Write a MATLAB program to compute the terms of a best-fit limaçoid.

The limaçoid cylinder parameters can be evaluated as shown below.

```
% M is a matrix of size n1*n2 x 3. n1 is the number of points
% per trace. n2 is the number of traces. The first column
% contains angle data (in radians), second column contains
% z height and third column has radial deviations
function [r, a, b, e, f] = limacoid(M,n1,n2)
n = size(M,1)
for i=1:n
    A(i,1) = cos(M(i,1));
    A(i,2) = sin(M(i,1));
    A(i,3) = M(i,2)*cos(M(i,1));
    A(i,4) = M(i,2)*sin(M(i,1));
    A(i,5) = 1;
    B(i,1) = M(i,3);
end
P = A\B;
a = P(1);
b = P(2);
e = P(3);
f = P(4);
r = P(5);
```

18.5 Errors in the Limaçon Approximation

In Sect. 18.2, we saw that the true measured radial deviation is given by Eq. 18.1, while the approximation is given by Eq. 18.4. In the following example, we evaluate the magnitude of error in the linear approximation. See Chetwynd and Phillipson (1980) for an excellent treatment of errors as a function of eccentricity to radius ratio. The ASME B89.3.1 (American Society of Mechanical Engineers 1972) provides guidelines on acceptable eccentricity to radius ratio.

Example 18.4. Plot the error (maximum radial error) in the limaçon approximation as a function of part radius, assuming part radius increases from 10 μm to 1 mm, while the eccentricity remains constant at 2 μm along 45° with the *x* axis.

The radial error in the limaçon approximation is the maximum difference between Eq. 18.1 and Eq. 18.4. Figure 18.2 shows that the radial errors in the approximation fall off rapidly as part diameter increases, for a given constant eccentricity.

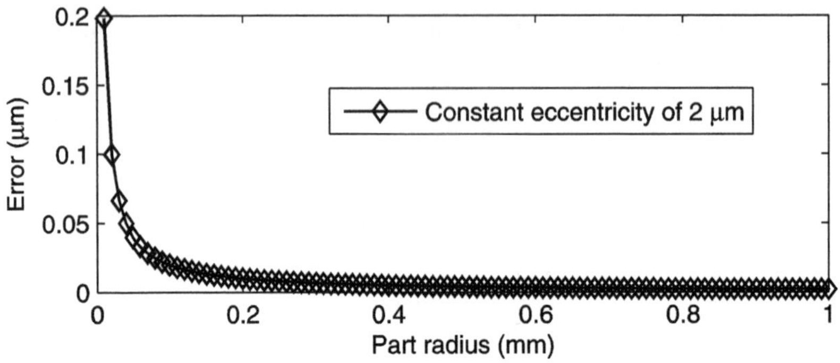

Fig. 18.2 Errors in limaçon approximation as function of part radius, but in the presence of constant eccentricity

```
% Generate the errors in limaçon approximation
theta = (0:1:360-1)*pi/180;
e = 0.002;
phi = pi/4;
i = 1;
for r = 0.01:0.01:1
    E = sqrt(r^2-(e^2).*sin(theta-phi)) - r;
    Err(i) = max(E);
    i = i+1;
end
plot(0.01:0.01:1,Err,'rd-');
```

18.6 Summary

During roundness measurement, it is not possible to center the part on the spindle precisely. Therefore, the radial deviations as measured by the instrument do not directly yield the desired out-of-roundness of the part. In Chap. 17, we reviewed non-linear LS techniques to determine the parameters of the best-fit circle. The deviations as measured from this best-fit circle yield the out-of-roundness, but the iterative method is quite cumbersome to implement. The limaçon approximation to a circle is valid when the eccentricity is small in comparison to the part's radius. The advantage of the limaçon approximation is that it is linear in its parameters, and provides the best-fit parameters with limited computational burden. Also, because roundness data is intrinsically collected in polar coordinates, the limaçon approximation is ideal because it too is formulated in polar coordinates, unlike the non-linear LS circle fitting routine, which is formulated in Cartesian coordinates. A further point to note is that this technique can also be used for partial arc fitting, as shown by Whitehouse (Whitehouse, 1973).

Exercises

18.1. Assume that a circular part of radius 3 mm is sampled every 0.1° on a roundness instrument. Further it is mounted with an eccentricity of 2 μm along the 45° line with the x axis. Also assume that the part has a 2 UPR signal of 1 μm amplitude. Let the probe reading at the start of the measurement be 20 μm. Generate the profile using the limaçon approximation and plot on a polar chart.

18.2. Determine the best-fit center and radius of the profile in Exercise 18.1 using (a) any non-linear LS method, (b) limaçon approximation. Explain why there is a small difference in the parameters.

18.3. Write a program to compute the distances of every point from a best-fit cylinder. The function should take the data point array and the parameters of the best-fit cylinder as arguments.

18.4. Assume that a cylindrical part of radius 3 mm is sampled every 1°, and at five different but equally spaced z heights on a roundness instrument. Assume that it has an eccentricity of $(2\,\mu\text{m}, 0)$ at $z = 0$, and $(0, 5\,\mu\text{m})$ at $z = 4$ mm. Generate the data as sampled by the instrument using the limaçon approximation.

18.5. Determine the best-fit parameters of the data in Exercise 18.4 using (a) any non-linear LS method, (b) limaçon approximation and verify that the obtained parameters are equal to those expected. If there is any discrepancy, explain the source.

References

American Society of Mechanical Engineers 1972, *ASME B89.3.1-1972, Measurement of Out-of-Roundness*, ASME. New York, NY, USA.

Chetwynd, D.G. 1979, 'Roundness measurement using limaçons', *Precision Engineering*, vol. 1, pp. 137–141.

Chetwynd, D.G. and Phillipson, P.H. 1980, 'An investigation of reference criteria used in roundness measurement', *Journal of Physics E: Scientific Instruments*, vol. 13, pp. 530–538.

Kakino, Y. and Kitazawa, J. 1978, 'In situ measurement of cylindricity', *Annals of the CIRP*, vol. 27, no. 1, pp. 371–375.

Whitehouse, D.J. 1973, 'A best-fit reference line for use in partial arcs', *Journal of Physics E: Scientific Instruments*, vol. 6, pp. 921–924.

Chapter 19
Exchange Algorithms for Minimum Zone

19.1 Introduction

The focus of this chapter is on exchange algorithms. We first review exchange algorithms for lines and planes, and then extend the argument for circles. We will discuss linear programming optimization-based techniques in Chap. 20 that can be used to determine the minimum-zone (MZ) limaçoid (cylinder) based on the limaçon approximation. We will then show in Chap. 20 that the linear programming based method provides an exchange algorithm for minimum-circumscribed and maximum-inscribed circles.

19.2 Exchange Algorithms

Exchange algorithms are one of the simplest and most intuitive approaches to solving MZ problems. They guarantee the smallest zone width (under certain conditions which we will discuss later), but are not applicable to all geometries. The general idea of an exchange algorithm is as follows

1. A random set of points are selected to define the MZ boundary for the first iteration. Three points are necessary for lines, four for planes and circles, *etc.* We refer to these points as the critical points because they define the zone.
2. We determine the MZ boundary for this initial estimate. This is a simple problem of fitting two parallel lines to three points, or two concentric circles to four points, *etc.*
3. Then, we identify the point that lies farthest away from either the lower (or inner) boundary or the upper (or outer) boundary. We refer to this point as the outlier point. If there are no outlier points, the algorithm has terminated. Otherwise we proceed to step 4.

4. We replace one of the current critical points by the outlier point. The problem of determining which critical point to replace is determined by brute force for some geometries. For others, we can devise efficient algorithms.
5. We go back to step 2 and determine the zone for the current set of critical points.

19.3 Exchange Algorithm for Line

Consider a line defined by

$$z = a + bx\,. \tag{19.1}$$

According to the definition of the MZ, all points should lie within two parallel lines given by

$$\begin{aligned} z &= a + h + bx \\ z &= a - h + bx \end{aligned}, \tag{19.2}$$

so that $h/\sqrt{1+b^2}$ is a minimum. We refer to the line defined in Eq. 19.1 as the mean line. The zone half-width is given by h (see Fig. 19.1a).

In order to proceed with the exchange algorithm, we arbitrarily select three points A, B, and C as the critical points of the first iteration. Assuming A and C lie on the upper line, and B on the lower line, we determine parameters a, b, and h by solving the system of equations:

$$\begin{aligned} z(A) &= a + h + bx(A) \\ z(B) &= a - h + bx(B) \\ z(C) &= a + h + bx(C) \end{aligned}\,. \tag{19.3}$$

We next determine the point that lies farthest away from either the upper or the lower line along the normal direction (we are interested in minimizing the zone width as measured perpendicular to the line, not along the z axis). Let this point be P. We now have to replace one of A, B, or C by P, and re-compute the parameters

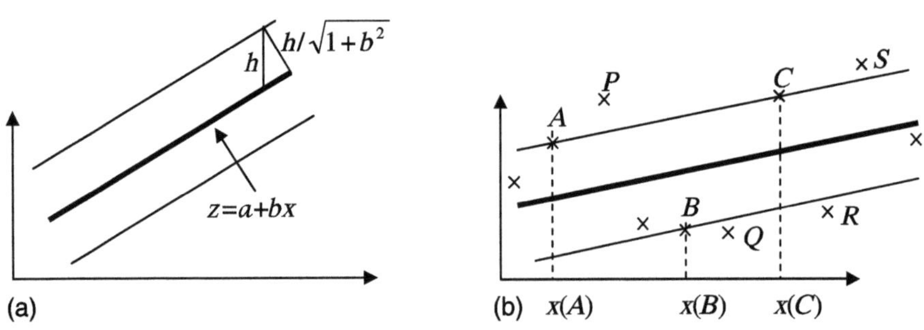

Fig. 19.1 **a** Line and its minimum-zone (MZ) boundaries. **b** Current critical points A, B, C, and outlier points P, Q, R, and S

of the line. A brute force approach involves sequentially replacing A, B, and C by P, and computing the zone parameters. That point, which when replaced by the outlier P, yields the smallest zone width is actually replaced for the next iteration. This method is computationally inefficient and is therefore never preferred. If an efficient algorithm is available to determine which point to replace, the exchange algorithm becomes invaluable. Luckily, we do have such an algorithm for the MZ line. We refer to this as the alternating rule and provide a motivation for its validity next.

The alternating rule states that two consecutive points that define the zone cannot lie on the same boundary line. Fukuda and Shimokohbe (1984) present an elegant proof of why this rule is valid. In Fig. 19.1b, the magnitude of the deviations of each of the three critical points (A, B, C) from the mean line is equal to h. Let the sign of the deviation at each of these points be given by the binary valued sign function sgn(). Then,

$$\begin{aligned} \operatorname{sgn}(A)h &= z(A) - bx(A) - a \\ \operatorname{sgn}(B)h &= z(B) - bx(B) - a \;. \\ \operatorname{sgn}(C)h &= z(C) - bx(C) - a \end{aligned} \tag{19.4}$$

For the arrangement of points in Fig. 19.1b, $\operatorname{sgn}(A) = 1$, $\operatorname{sgn}(B) = -1$, and $\operatorname{sgn}(C) = 1$. For this derivation, we assume that the signs of the deviations at A, B, and C are not known. Our only assumption is that the points are ordered, that is, $x(A) < x(B) < x(C)$. We can solve Eq. 19.4 for h as

$$h = \frac{z(A)(x(B)-x(C)) + z(B)(x(C)-x(A)) + z(C)(x(A)-x(B))}{\operatorname{sgn}(A)(x(B)-x(C)) + \operatorname{sgn}(B)(x(C)-x(A)) + \operatorname{sgn}(C)(x(A)-x(B))} \,. \tag{19.5}$$

From Eq. 19.5, we see that the minimum value for h occurs when

$$\begin{aligned} \operatorname{sgn}(A) &= \operatorname{sgn}(x(B)-x(C)) \\ \operatorname{sgn}(B) &= \operatorname{sgn}(x(C)-x(A)) \\ \operatorname{sgn}(C) &= \operatorname{sgn}(x(A)-x(B)) \,. \end{aligned} \tag{19.6}$$

Because the points are ordered, $x(B) - x(C) < 0$, $x(C) - x(A) > 0$, and $x(A) - x(B) < 0$. Therefore $\operatorname{sgn}(A) = -1$, $\operatorname{sgn}(B) = 1$, and $\operatorname{sgn}(C) = -1$ or the residuals alternate in sign (note that Eq. 19.5 may also be equivalently written by multiplying both the numerator and denominator by -1; this would result in $\operatorname{sgn}(A) = 1$, $\operatorname{sgn}(B) = -1$, and $\operatorname{sgn}(C) = 1$). Thus, no two consecutive points can lie on the same boundary line.

In Fig. 19.1b, if point A lies on the upper line, point B will lie on the lower and C on the upper line. Given the alternating rule, we can define rules to identify which of the three critical points to be replaced without having to determine the width of the zone for each of the three possible cases. Consider the different cases shown in Fig. 19.1b. Thus, if point P were the outlier, we replace point A by point P because the new resulting set of three points satisfies the alternating rule. If Q were the outlier instead of P, then, we would replace point B by point Q. If point S is the outlier, we replace point C. However, if point R is the outlier, we have to replace point *A* because no other combination of three points (*ABR* or *ACR*) will satisfy the alternating rule. In this case, we also have to reorder the points so that *BCR* is now the new set *ABC*.

Example 19.1. Given a set of data points on a line, write a program that computes the MZ line.

*Index*1, *Index*2 and *Index*3 are variables that store the indices of the current set of critical points ordered so that *Index*1 is the smallest and *Index*3 is the largest; z and x are arrays of deviations and locations. The program first fits the pair of MZ lines through the current set of points and finds the point that lies farthest away. The x position of this point is stored in b. Subsequently, the alternating rule is implemented where the point that needs to be discarded is determined.

```
function [slope, intercept1, intercept2, h] =
MZLINE(z, x, Index1, Index2, Index3)
flag = true;
eps = 1e-6;
t = 1;
while flag
    n = size(z,2);
    [P,S] = polyfit([x(Index1) x(Index3)],[z(Index1)
                     z(Index3)],1);
    slope = P(1);
    intercept1 = P(2);
    intercept2 = z(Index2) - slope*x(Index2);
    h = abs(intercept1-intercept2)/2;
    distance = z - slope.*x - (intercept1 + intercept2)/2;
    [a,b] = max(abs(distance));
    if abs(a-h) <= eps
        flag = false;
    else
        if b < x(Index1)
            if (z(b) - (slope*x(b)+(intercept1 + intercept2)/2))
*(z(Index1) - (slope*x(Index1)+(intercept1 + intercept2)/2)) > 0
                Index1 = b;
            else
                Index3 = b;
            end
        elseif b < x(Index2)
            if (z(b) - (slope*x(b)+(intercept1 + intercept2)/2))
*(z(Index2) - (slope*x(Index2)+(intercept1 + intercept2)/2)) > 0
                Index2 = b;
            else
                Index1 = b;
            end
        elseif b < x(Index3)
            if (z(b) - (slope*x(b)+(intercept1 + intercept2)/2))
*(z(Index3) - (slope*x(Index3)+(intercept1 + intercept2)/2)) > 0
                Index3 = b;
            else
                Index2 = b;
            end
```

```
        else
            if (z(b) - (slope*x(b)+(intercept1 + intercept2)/2))
*(z(Index3) - (slope*x(Index3)+(intercept1 + intercept2)/2)) > 0
                Index3 = b;
            else
                Index1 = b;
            end
        end
    end
    temp1 = sort([Index1 Index2 Index3]);
    Index1 = temp1(1);
    Index2 = temp1(2);
    Index3 = temp1(3);
    t = t+1;
end
[Index1 Index2 Index3 h t]
```

19.4 Exchange Algorithm for Plane

The alternating rule is not as straightforward in the case of planes because the points are not ordered. We can extend the argument of Fukuda and Shimokohbe (1984) for a plane of the form $z = ax + by + c$, to determine how the points have to be distributed on each plane. Let four points A, B, C, and D define the zone. Then, the residuals at each of the points are equal. We study the sign of the residuals as before. Let h be the magnitude of the half-width of the zone. Then,

$$\begin{aligned}
\operatorname{sgn}(A)h &= z(A) - ax(A) - by(A) - c \\
\operatorname{sgn}(B)h &= z(B) - ax(B) - by(B) - c \\
\operatorname{sgn}(C)h &= z(C) - ax(C) - by(C) - c \\
\operatorname{sgn}(D)h &= z(D) - ax(D) - by(D) - c
\end{aligned} . \tag{19.7}$$

This system of four equations can be solved for h resulting in

$$h = \frac{z(A)T_1 + z(B)T_2 + z(C)T_3 + z(D)T_4}{\operatorname{sgn}(A)T_1 + \operatorname{sgn}(B)T_2 + \operatorname{sgn}(C)T_3 + \operatorname{sgn}(D)T_4}, \tag{19.8}$$

where

$$\begin{aligned}
T_1 &= x_{BC}x_{DC}\left(\frac{y_{BC}}{x_{BC}} - \frac{y_{DC}}{x_{DC}}\right) \\
T_2 &= x_{DC}x_{AC}\left(\frac{y_{DC}}{x_{DC}} - \frac{y_{AC}}{x_{AC}}\right) \\
T_3 &= x_{AB}x_{DB}\left(\frac{y_{AB}}{x_{AB}} - \frac{y_{DB}}{x_{DB}}\right) \\
T_4 &= x_{CB}x_{AB}\left(\frac{y_{CB}}{x_{CB}} - \frac{y_{AB}}{x_{AB}}\right)
\end{aligned}, \tag{19.9}$$

and for any two points i and j, $x_{ij} = (x(i) - x(j)), y_{ij} = (y(i) - y(j))$.

The width of the zone is minimum when

$$\begin{aligned} \text{sgn}(A) &= \text{sgn}(T_1) \\ \text{sgn}(B) &= \text{sgn}(T_2) \\ \text{sgn}(C) &= \text{sgn}(T_3) \\ \text{sgn}(D) &= \text{sgn}(T_4) \end{aligned}. \tag{19.10}$$

For any orientation of points A, B, C, and D, we can examine the expressions for T_1, T_2, T_3, and T_4 to determine the relative signs of the residuals. We illustrate the process for two configurations, the argument for which is applicable to any configuration.

In Fig. 19.2, $T_1 > 0$ because $x_{BC} < 0, x_{DC} > 0, \frac{y_{BC}}{x_{BC}} < 0$, and $\frac{y_{DC}}{x_{DC}} > 0$. Therefore we set $\text{sgn}(A) = 1$. Next, consider the expression T_2. $T_2 < 0$ because $\frac{y_{DC}}{x_{DC}} - \frac{y_{AC}}{x_{AC}} > 0$ (point A is in the half-plane defined by line DC containing point B, and therefore the magnitude of the slope of AC is smaller than that of DC), $x_{DC} > 0$ and $x_{AC} < 0$. Therefore we set $\text{sgn}(B) = -1$. Using similar arguments, we can show that $T_3 > 0$ and $T_4 < 0$, resulting in $\text{sgn}(C) = 1$ and $\text{sgn}(D) = -1$. When A is in the half plane defined by line DC, but not containing point B, we can show that $T_1 > 0$, $T_2 > 0$, $T_3 < 0$, and $T_4 > 0$, which means that $\text{sgn}(A) = \text{sgn}(B) = \text{sgn}(D) = 1$, and $\text{sgn}(C) = -1$.

This argument can be generalized for any configuration of points. By working out different scenarios, it can be seen that there are two possible solution cases for MZ, which form the alternating rule. First, the projection of the four points on the xy plane forms a convex quadrilateral. In such a situation, one pair of diagonal points lies on one minimum-zone plane while the other pair lies on the other plane. The second possibility is that one minimum-zone plane contacts one point while the other plane contacts three points. In this case, the projection of the four points on the xy plane forms a triangle with one point (the single point that contacts a plane) inside the triangle (defined by three points on the other plane). These situations are illustrated in Fig. 19.2. The results are summarized in Chetwynd (1985), but he arrives at the conclusion from a linear programming perspective.

On the basis of this alternating rule, it is possible to develop an algorithm that will determine which of the critical points to replace by the outlier point at each

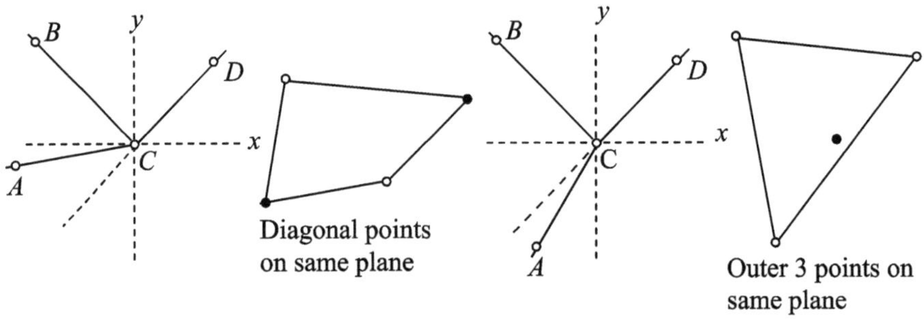

Fig. 19.2 Illustration of exchange algorithm for planes (projection of points on xy plane shown)

iteration. Of course, this involves examining a variety of cases and it may not in fact be any more efficient than the brute force method! We have illustrated the exchange algorithm for planes to highlight the fact that it is not always as efficient as it sounds.

19.5 Exchange Algorithm for Circle

The alternating rule says that the points that define the two circles are to be distributed so that no two consecutive points shall lie on the same circle. Thus, if the zone is currently defined by points *ABCD* in that order (starting at positive *x* axis and going counter-clockwise), and if *A* lies on the inner circle as shown in Fig. 19.3, point *B* has to lie on the outer circle, point *C* on the inner and point *D* on the outer. As in the case of the line, we can develop an algorithm that determines which of the critical points to replace with the outlier. Consider point *P* as the outlier. Because *P* lies inside of the inner circle, we can only consider points that lie on the inner circle as valid candidates (*A* and *C*) to be replaced. Between *A* and *C*, we can only replace point *A* because replacing point *C* will violate the alternating rule (the new set would then be *PABD*: *P* and *A* lie on the same circle). Consider other cases for outliers: if *Q* is the outlier, we replace *B*, if *R* is the outlier, we again replace *B*, if *S* is the outlier, we replace *D*, if *T* is the outlier, we replace *C*, if *U* is the outlier, we replace *A* and re-order so that the sequence *BCDU* is the new *ABCD*.

Following the argument of Fukuda and Shimokohbe (1984), we demonstrate here the alternating rule for circles. Let four points *A*, *B*, *C*, and *D* define the zone. Then, the residuals at each of the points are equal. We study the sign of the residuals as before. Let *h* be the magnitude of the half-width of the zone. Then

$$\begin{aligned} \operatorname{sgn}(A)h &= R(A) - a\cos\theta(A) - b\sin\theta(A) - r \\ \operatorname{sgn}(B)h &= R(B) - a\cos\theta(B) - b\sin\theta(B) - r \\ \operatorname{sgn}(C)h &= R(C) - a\cos\theta(C) - b\sin\theta(C) - r \\ \operatorname{sgn}(D)h &= R(D) - a\cos\theta(D) - b\sin\theta(D) - r \end{aligned}, \qquad (19.11)$$

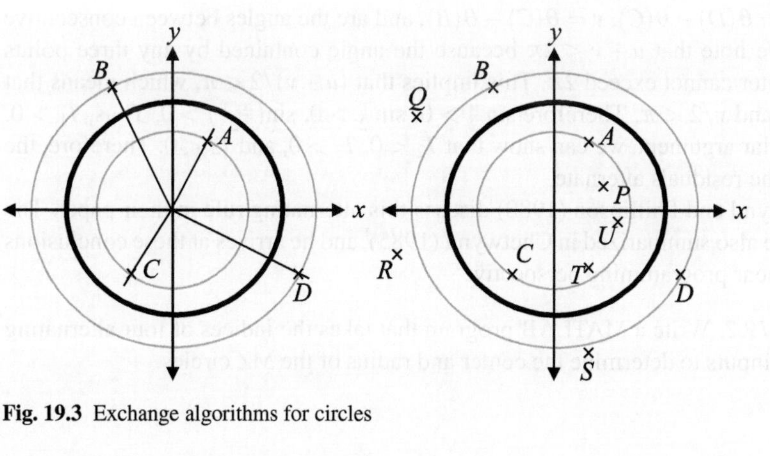

Fig. 19.3 Exchange algorithms for circles

where $R(i)$ is the radial distance of the ith point, and (a,b) and r are the center and radius of the mean circle. This system of four equations can be solved for h resulting in

$$h = k\frac{R(A)T_1 + R(B)T_2 + R(C)T_3 + R(D)T_4}{\text{sgn}(A)T_1 + \text{sgn}(B)T_2 + \text{sgn}(C)T_3 + \text{sgn}(D)T_4}, \tag{19.12}$$

where $k = \sin(\theta(B)) - \sin(\theta(C))$, and

$$\begin{aligned} T_1 &= \sin(\theta(D) - \theta(C)) + \sin(\theta(C) - \theta(B)) + \sin(\theta(B) - \theta(D)) \\ T_2 &= \sin(\theta(A) - \theta(C)) + \sin(\theta(C) - \theta(D)) + \sin(\theta(D) - \theta(A)) \\ T_3 &= \sin(\theta(B) - \theta(A)) + \sin(\theta(D) - \theta(B)) + \sin(\theta(A) - \theta(D)) \\ T_4 &= \sin(\theta(A) - \theta(B)) + \sin(\theta(B) - \theta(C)) + \sin(\theta(C) - \theta(A)) \end{aligned}. \tag{19.13}$$

The width of the zone is minimum when

$$\begin{aligned} \text{sgn}(A) &= \text{sgn}(T_1) \\ \text{sgn}(B) &= \text{sgn}(T_2) \\ \text{sgn}(C) &= \text{sgn}(T_3) \\ \text{sgn}(D) &= \text{sgn}(T_4) \end{aligned}. \tag{19.14}$$

For ordered points A, B, C, and D $(0 < \theta(A) < \theta(B) < \theta(C) < \theta(D) < 2\pi)$, we examine the expressions T_1, T_2, T_3, and T_4 to determine the relative signs of the residuals.

Consider the expression T_1. It can be expressed as

$$\begin{aligned} T_1 &= \sin(\theta(D) - \theta(C)) + \sin(\theta(C) - \theta(B)) + \sin(\theta(B) - \theta(D)) \\ &= 4\sin\frac{u}{2}\sin\frac{v}{2}\sin\left(\frac{u+v}{2}\right), \end{aligned} \tag{19.15}$$

where $u = \theta(D) - \theta(C)$, $v = \theta(C) - \theta(B)$, and are the angles between consecutive points. We note that $u + v < 2\pi$ because the angle contained by any three points at the center cannot exceed 2π. This implies that $(u+v)/2 < \pi$, which means that $u/2 < \pi$ and $v/2 < \pi$. Therefore $\sin\frac{u}{2} > 0$, $\sin\frac{v}{2} > 0$, $\sin(\frac{u+v}{2}) > 0$. Thus, $T_1 > 0$. By a similar argument, we can show that $T_2 < 0$, $T_3 > 0$, and $T_4 < 0$. Therefore, the signs of the residuals alternate.

Chetwynd and Phillipson (1980) discuss this alternating rule in their paper. The results are also summarized in Chetwynd (1985), and he arrives at these conclusions from a linear programming perspective.

Example 19.2. Write a MATLAB program that takes the indices of four alternating points as inputs to determine the center and radius of the MZ circles.

R is the array of radial deviations, p_1, p_2, p_3 and p_4 are the indices of the current set of critical points and *theta* is an array of angular location of the points. We assume that the first and the third point are on the outer circle, hence the positive coefficient for *h*.

```
function [r,h,a,b] = limaconfitMZ(R,p1,p2,p3,p4,theta)
A=[1 1 cos(theta(p1)) sin(theta(p1));
    1 -1 cos(theta(p2)) sin(theta(p2));
    1 1 cos(theta(p3)) sin(theta(p3));
    1 -1 cos(theta(p4)) sin(theta(p4))];
B=[R(p1); R(p2); R(p3); R(p4)];
P = inv(A)*B;
r = P(1);
h = P(2);
a = P(3);
b = P(4);
```

Example 19.3. Write a program that implements the alternating rule for circles.

The indices of the current set of critical points are p_1, p_2, p_3 and p_4, *index* is the index of the point that lies farthest away from the circle, *deviation* is the radial deviation of the point that lies farthest away from the circle and *h* is the current zone half-width. Note that *h* can be positive or negative; if the assumption that the first and third points are on the outer circle is incorrect, *h* is negative.

```
function [p1,p2,p3,p4] =
alternatingrule(p1,p2,p3,p4,index,h,deviation)
[p1 p2 p3 p4 index h deviation]
if (index < p1)
    if (deviation >0 & h>0) | (deviation < 0 & h<0)
        p1 = index;
    else % (deviation >0 & h<0) | (deviation <0 & h>0)
        p4=p3;p3=p2;p2=p1;p1=index;
    end
elseif (index < p2)
    if (deviation >0 & h>0) | (deviation < 0 & h<0)
        p1 = index;
    else % (deviation >0 & h<0) | (deviation <0 & h>0)
        p2 = index;
    end
elseif (index < p3)
    if (deviation >0 & h>0) | (deviation < 0 & h<0)
        p3 = index;
    else % (deviation >0 & h<0) | (deviation <0 & h>0)
        p2 = index;
    end
elseif (index < p4)
    if (deviation >0 & h>0) | (deviation < 0 & h<0)
        p3 = index;
    else % (deviation >0 & h<0) | (deviation <0 & h>0)
        p4 = index;
    end
else
```

```
    if (deviation >0 & h>0) | (deviation < 0 & h<0)
        p1=p2;p2=p3;p3=p4;p4=index;
    else % (deviation >0 & h<0) | (deviation <0 & h>0)
        p4 = index;
    end
end
```

Example 19.4. Write a program that implements the MZ circle.

```
function [r, a, b, h, OOR] = mzcircle(R)
n = size(R,1);
theta= (1:1:n)';
theta = (theta-1)*2*pi/n;
p1 = 1;
p2 = round(3*n/8);
p3 = round(5*n/8);
p4 = round(7*n/8);
[r,h,a,b] = limaconfitMZ(R,p1,p2,p3,p4,theta);
proceed = 1;
while proceed == 1
    temp = abs(h);
    index = -1;
    for i=1:n
        deviation = R(i)-r-a*cos(theta(i))-b*sin(theta(i));
        if (abs(deviation) > temp+1e-4)
            temp = abs(deviation);
            index = i;
        end
    end
    if (index == -1 | index == p1 | index == p2 |
index == p3 | index == p4)
        proceed = 0;
    else
        deviation = R(index)-(r+a*cos(theta(index))+b*
sin(theta(index)));
        [p1,p2,p3,p4] = alternatingrule(p1,p2,p3,p4,
index,h,deviation);
        [r,h,a,b] = limaconfitMZ(R,p1,p2,p3,p4,theta);
    end
end
R_MZ = r+a*cos(theta)+b*sin(theta);
OOR = 2*h
[r a b h 2*h]
```

19.6 Summary

Exchange algorithms are simple to implement, but have limited applicability. We have discussed the application of exchange algorithm for lines, planes, and circles in this chapter. We show how the alternating rule is beneficial in the case of lines and circles, but not in the case of planes.

Exercises

19.1. Write a MATLAB program that computes the MZ line using the brute force method.

19.2. For the data below, compute the MZ line using both the brute force method and using the alternating rule. How many exchanges take place in each case?

x	1	2	3	4	5	6	7
z	4.7963	4.96	2.8753	0.6094	2.6247	1.8626	9.1709
x	8	9	10	11	12	13	14
z	1.2328	0.1345	3.6969	6.9864	8.8935	5.9377	1.5669
x	15	16	17	18	19	20	21
z	3.1669	2.334	0.0842	3.969	6.4987	0.85	7.6881

All units in micrometers.

19.3. Simulate straightness data obtained from a nominally flat part assuming 40 points are sampled at 100 μm spacing, and that the part has a sinusoidal form of amplitude 100 μm and wavelength 2 mm. For this part, compute the MZ line parameters using (a) the alternating rule and (b) the brute force method. Compare the time taken for each method.

19.4. For the roundness data in Example 7.3, determine the parameters of the MZ circle using the alternating rule.

References

Chetwynd, D.G. 1985, 'Applications of linear programming to engineering metrology', *Proceedings of the Institution of Mechanical Engineers Part B: Management and Engineering Manufacture*, vol. 199, no. B2, pp. 93–100.

Chetwynd, D.G. and Phillipson, P.H. 1980, 'An investigation of reference criteria used in roundness measurement', *Journal of Physics E: Scientific Instruments*, vol. 13, pp. 530–538.

Fukuda, M. and Shimokohbe, A. 1984, 'Algorithms for form error evaluation-methods of the minimum zone and the least squares', *Proceedings of the International Symposium on Metrology for Quality Control in Production, Tokyo*, pp. 197–202.

Chapter 20
Reference Circle-Fitting Using Linear Programming Simplex

20.1 Introduction

The minimum-zone (MZ), the minimum-circumscribed (MC), and the maximum-inscribed (MI) geometry determination problems are essentially optimization problems. These are challenging to solve when the objective function and/or the constraints are non-linear functions of the unknown parameters. In roundness metrology, we commonly use linear approximations of the circle and cylinder. Because the objective function and the constraints are linear functions of the unknowns, the MZ, MC, and MI circle (and cylinder) problems can be conveniently solved using the linear programming (LP) simplex technique. We discuss the application of the simplex method for MI and MC circles in this chapter. They can be easily extended for MZ circles, as well as for MZ cylinders.

The method discussed here is closely linked to exchange algorithms, such as the ones discussed in Chap. 19. We demonstrate this link by extending our discussion on LP methods to develop an exchange algorithm for MI and MC problems. All techniques discussed here were first reported by Chetwynd (1985); his early paper on LP applications in metrology is a definitive reference. The material presented here is an adaptation of that work for introductory readers. Chetwynd's original publication is cited in the references and also listed below in the footnote with publisher information as requested by the publisher[1].

20.2 LP Simplex

A LP problem is an optimization of an objective function that is a linear function of some variables in the presence of constraints that are also linear functions of

[1] Proceedings of the Institution of Mechanical Engineers, Part B: Management and Engineering Manufacture, 199 (B2), pp. 93–100, 1985, Professional Engineering Publishing, ISSN: 0954-4054.

those variables. A standard LP problem with two variables and n constraints is of the form:

$$\begin{aligned} \text{Maximize} \quad & E = ax + by \\ \text{such that} \quad & c(i)x + d(i)y \leqslant e(i) \,, \quad \text{for} \quad i = 1:n \,. \end{aligned} \tag{20.1}$$

The constraints, all of which are of the less-than-or-equal-to type, form a region in space that is referred to as a convex polytope; for the case with two variables, the region defined by the constraints can be considered a convex polygon. Standard LP formulation has the following properties: it involves a maximization, not minimization, of the objective function; all unknown variables (x, y) are non-negative (greater than or equal to zero); and the right hand side $e(i)$ is also non-negative. The origin is therefore a feasible solution to the problem. That is, it satisfies all constraints including the non-negativity constraint on the variables. Having all of the constraints as less-than-or-equal-to type, and the origin as a starting feasible solution is computationally advantageous.

The simplex algorithm starts at the origin and then proceeds from one vertex to another until it finds the optimal solution. It can be shown that if an optimal solution exists, it will occur at one of the vertices of the defining boundary, and not at the interior.

Before the simplex algorithm can be initiated however, each of the less-than-or-equal-to constraints is converted to more manageable equal-to type by adding non-negative slack variables $s(i)$. Therefore, we now have a system of n equations and $n+2$ variables

$$\begin{aligned} \text{Maximize} \quad & E = ax + by \\ \text{such that} \quad & c(i)x + d(i)y + s(i) = e(i) \,, \quad \text{for} \quad i = 1:n \,, \\ & x, y, s(i) \geqslant 0 \,, \quad \text{for} \quad i = 1:n \,. \end{aligned} \tag{20.2}$$

The origin, as we mentioned above, is a feasible solution and is the starting solution for the simplex algorithm. This implies that the n slack variables $s(i)$ take on as large a value as possible, that is, $s(i) = e(i)$. Note that the objective function is only dependent on x and y, and therefore, as the simplex proceeds, it identifies slack variables whose value can be reduced (preferably to zero) while the values of x and y are increased until such point where no further changes are possible. Geometrically, it achieves this by hopping from one vertex of the defining perimeter to the next.

It is impossible to provide a comprehensive description of the technique here; we refer the readers not familiar with this subject to numerous excellent text books, for instance (Sultan 1993; Ignizio 1981) on this topic. In Sect. 20.3, we illustrate the working of the simplex algorithm in the context of a MI circle problem. Note that for convenience we use the term circle to represent its linear approximation, which is the limaçon.

20.3 Formulating the MI Problem

The MI circle is the circle of largest radius that can be constructed so that all measured data points reside on or outside the circle. Mathematically, the problem can be formulated as:

$$\begin{aligned} &\text{Maximize} \quad E = r \\ &\text{such that:} \quad a\cos\theta(i) + b\sin\theta(i) + r \leqslant R(i)\,, \quad \text{for} \quad i = 1:n\,, \end{aligned} \tag{20.3}$$

where $a\cos\theta(i) + b\sin\theta(i) + r$ is the ith point on the MI circle, (a, b) is the center and r is the radius of the MI circle. $R(i)$ is the measured radial form on the part at $\theta(i)$.

This problem has three variables and n constraints. As mentioned before, LP formulation requires that the unknown parameters be non-negative. While that is true for the radius r, it is not necessary for the center (a,b). In order to allow the center to take negative values, we express a and b each as a difference of two non-negative parameters. This is a standard LP technique. In order to proceed with the simplex algorithm, we convert each of the less-than-or-equal-to constraints to equal-to type by adding non-negative slack variables $s(i)$. The MI circle problem is therefore stated as:

$$\begin{aligned} &\text{Maximize} \quad E = r\,, \quad \text{such that} \\ &(a_1 - a_2)\cos\theta(i) + (b_1 - b_2)\sin\theta(i) + r + s(i) = R(i)\,, \quad \text{for} \quad i = 1:n \\ &a_1 \geqslant 0\,,\; a_2 \geqslant 0\,,\; b_1 \geqslant 0\,,\; b_2 \geqslant 0\,,\; s(i) \geqslant 0\,,\; r \geqslant 0\,, \quad \text{for} \quad i = 1:n\,. \end{aligned} \tag{20.4}$$

Because all measured $R(i)$ are non-negative (if they are not, they can be forced to become non-negative by adding a positive constant), $(a_1, b_1, a_2, b_2, r) = (0,0,0,0,0)$ and $s(i) = R(i)$ is a starting basic feasible solution. That is, our starting solution is a circle centered at the origin and of zero radius, which is one possible inscribed circle. The MI problem can now be solved using the simplex method, which will attempt to gradually move the center and increase the radius to a point where no further perturbations of the parameters will yield an increased radius. We illustrate the working of the simplex algorithm using the following example.

Example 20.1. Determine the MI circle using the LP simplex method for the data below (given in polar form).

Angle (rad)	0	0.7854	1.5708	2.3562	3.1416	3.9270	4.7124	5.4978
Radius (μm)	11	14	10	16	13	12	15	14

The system of equations can be expressed in tableau form as shown in Table 20.1. Each row represents a constraint, except for the last row, which represents the objective function. The values in the table are the coefficients of the unknown parameters. Thus, constraint #3 reads:

$$\begin{aligned} &0.a_1 - 0.a_2 + 1.b_1 - 1.b_2 + r \\ &+ 0.s(1) + 0.s(2) + 1.s(3) + 0.s(4) + 0.s(5) + 0.s(6) + 0.s(7) + 0.s(8) = 10 \end{aligned}$$

Table 20.1 Starting tableau

	a_1	a_2	b_1	b_2	r	$s(1)$	$s(2)$	$s(3)$	$s(4)$	$s(5)$	$s(6)$	$s(7)$	$s(8)$	R	R/r
1	1.00	−1.00	0.00	0.00	1.00	1.00	0.00	0.00	0.00	0.00	0.00	0.00	0.00	11.00	11.00
2	0.71	−0.71	0.71	−0.71	1.00	0.00	1.00	0.00	0.00	0.00	0.00	0.00	0.00	14.00	14.00
3	0.00	0.00	1.00	−1.00	1.00	0.00	0.00	1.00	0.00	0.00	0.00	0.00	0.00	10.00	10.00
4	−0.71	0.71	0.71	−0.71	1.00	0.00	0.00	0.00	1.00	0.00	0.00	0.00	0.00	16.00	16.00
5	−1.00	1.00	0.00	0.00	1.00	0.00	0.00	0.00	0.00	1.00	0.00	0.00	0.00	13.00	13.00
6	−0.71	0.71	−0.71	0.71	1.00	0.00	0.00	0.00	0.00	0.00	1.00	0.00	0.00	12.00	12.00
7	0.00	0.00	−1.00	1.00	1.00	0.00	0.00	0.00	0.00	0.00	0.00	1.00	0.00	15.00	15.00
8	0.71	−0.71	−0.71	0.71	1.00	0.00	0.00	0.00	0.00	0.00	0.00	0.00	1.00	14.00	14.00
E	0.00	0.00	0.00	0.00	1.00	0.00	0.00	0.00	0.00	0.00	0.00	0.00	0.00	0.00	

```
% A is a matrix of coefficients of the constraint; F is an array
% of coefficients of the objective function
theta = (0:7)'/8*2*pi;
A = [cos(theta) -cos(theta) sin(theta) -sin(theta) ones(8,1)
diag(ones(8,1))];
F = [0;0;0;0;1;zeros(8,1);0]';
R = [1.1;1.4;1;1.6;1.3;1.2;1.5;1.4]*10;
Z = [A R];
[Z;F]
```

Our starting solution is $(a_1,b_1,a_2,b_2,r) = (0,0,0,0,0)$ and $s(i) = R(i)$. Because (a_1,b_1,a_2,b_2,r) are all zero, we refer to these as the non-basic variables. Each of the n non-zero $s(i)$ are referred to as basic variables.

The only possible way of increasing the objective function value at this time is to increase the radius. This is evident from the last row of the table where the only non-negative coefficient occurs for r, that is, $E = 1.r$. The radius however cannot be increased infinitely; constraint #3 limits the increase to a value of 10.

We therefore express constraint #3 as

$$r = 10 - [0.a_1 - 0a_2 + 1.b_1 - 1.b_2 + 0.s(1) + 0.s(2) + 1.s(3) + 0.s(4) \\ + 0.s(5) + 0.s(6) + 0.s(7) + 0.s(8)]$$

and substitute this value of r in all the other constraints. After this substitution, we express the system of equations with all unknown parameters in the left hand side and constant term on the right hand side. Notice that constraint #3 does not appear to have changed; however, if r had a coefficient other than 1, it will now be 1 as a result of this transformation. The table after the transformation is shown next. What we have effectively accomplished, in LP terminology, is to bring r into while removing $s(3)$ from the basis.

```
% Basic variables are s(i) (for i = 1,2,3,4,5,6,7,8) and nonbasic
% variables are (a1,a2,b1,b2,r)
% Bring r into basic -- s(3) becomes nonbasic
ColNo = 5; rowNo = 3;
Z(rowNo,:) = Z(rowNo,:)/ Z(rowNo,ColNo);
Z1 = Z;
for i=1:8
    if i ~= rowNo
        Z1(i,:) = Z(i,:) - Z(i,ColNo)*Z(rowNo,:);
```

Table 20.2 Tableau after one iteration of the simplex

	a_1	a_2	b_1	b_2	r	$s(1)$	$s(2)$	$s(3)$	$s(4)$	$s(5)$	$s(6)$	$s(7)$	$s(8)$	R	R/b_2
1	1.00	−1.00	−1.00	1.00	0.00	1.00	0.00	−1.00	0.00	0.00	0.00	0.00	0.00	1.00	1.00
2	0.71	−0.71	−0.29	0.29	0.00	0.00	1.00	−1.00	0.00	0.00	0.00	0.00	0.00	4.00	13.66
3	0.00	0.00	1.00	−1.00	1.00	0.00	0.00	1.00	0.00	0.00	0.00	0.00	0.00	10.00	−10.00
4	−0.71	0.71	−0.29	0.29	0.00	0.00	0.00	−1.00	1.00	0.00	0.00	0.00	0.00	6.00	20.49
5	−1.00	1.00	−1.00	1.00	0.00	0.00	0.00	−1.00	0.00	1.00	0.00	0.00	0.00	3.00	3.00
6	−0.71	0.71	−1.71	1.71	0.00	0.00	0.00	−1.00	0.00	0.00	1.00	0.00	0.00	2.00	1.17
7	0.00	0.00	−2.00	2.00	0.00	0.00	0.00	−1.00	0.00	0.00	0.00	1.00	0.00	5.00	2.50
8	0.71	−0.71	−1.71	1.71	0.00	0.00	0.00	−1.00	0.00	0.00	0.00	0.00	1.00	4.00	2.34
E	0.00	0.00	−1.00	1.00	0.00	0.00	0.00	−1.00	0.00	0.00	0.00	0.00	0.00	−10.00	

```
    end
end
F1 = F - F(ColNo)*Z(rowNo,:);
[Z1;F1]
```

The largest increase in the objective function can now be obtained by increasing b_2 (from the last row). Constraint #1 limits the extent to which b_2 can be increased. Therefore, bringing b_2 into the basis while removing $s(1)$ from the basis (which means expressing constraint #1 in terms of b_2) can be accomplished as shown below:

```
% Basic variables are r,s(i), i = 1,2,4,5,6,7,8.
% Nonbasic variables are a1,a2,b1,b2,s(3)
% bring b2 into the basis and s(1) out of basis
ColNo = 4; rowNo = 1;
Z1(rowNo,:) = Z1(rowNo,:)/ Z1(rowNo,ColNo);
Z2 = Z1;
for i=1:8
    if i ~= rowNo
        Z2(i,:) = Z1(i,:) - Z1(i,ColNo)*Z1(rowNo,:);
    end
end
F2 = F1 - F1(ColNo)*Z1(rowNo,:);
[Z2; F2]
```

The largest increase in the objective function can now be obtained by increasing a_2. Constraint #6 limits the extent to which a_2 can be increased.

```
% Basic variables are r,b2, s(i), i = 2,4,5,6,7,8.
% Nonbasic variables are a1,a2,b1,s(i), i = 1,3.
% bring a2 into basis and s(6) out
ColNo = 2; rowNo = 6;
Z2(rowNo,:) = Z2(rowNo,:)/ Z2(rowNo,ColNo);
Z3 = Z2;
for i=1:8
    if i ~= rowNo
        Z3(i,:) = Z2(i,:) - Z2(i,ColNo)*Z2(rowNo,:);
    end
end
F3 = F2 - F2(ColNo)*Z2(rowNo,:);
[Z3;F3]
```

Table 20.3 Tableau after two iterations of the simplex

	a_1	a_2	b_1	b_2	r	$s(1)$	$s(2)$	$s(3)$	$s(4)$	$s(5)$	$s(6)$	$s(7)$	$s(8)$	R	R/b_2
1	1.00	−1.00	−1.00	1.00	0.00	1.00	0.00	−1.00	0.00	0.00	0.00	0.00	0.00	1.00	−1.00
2	0.41	−0.41	0.00	0.00	0.00	−0.29	1.00	−0.71	0.00	0.00	0.00	0.00	0.00	3.71	−8.95
3	1.00	−1.00	0.00	0.00	1.00	1.00	0.00	0.00	0.00	0.00	0.00	0.00	0.00	11.00	−11.00
4	−1.00	1.00	0.00	0.00	0.00	−0.29	0.00	−0.71	1.00	0.00	0.00	0.00	0.00	5.71	5.71
5	−2.00	2.00	0.00	0.00	0.00	−1.00	0.00	0.00	0.00	1.00	0.00	0.00	0.00	2.00	1.00
6	−1.00	1.00	0.00	0.00	0.00	−0.71	0.00	0.29	0.00	0.00	0.41	0.00	0.00	0.12	0.12
7	−2.00	2.00	0.00	0.00	0.00	−2.00	0.00	1.00	0.00	0.00	0.00	1.00	0.00	3.00	1.50
8	−1.00	1.00	0.00	0.00	0.00	−1.71	0.00	0.71	0.00	0.00	0.00	0.00	1.00	2.29	2.29
E	−1.00	1.00	0.00	0.00	0.00	−1.00	0.00	0.00	0.00	0.00	0.00	0.00	0.00	−11.00	

The objective function cannot be increased any further, and we have therefore reached optimality. From Table 20.4, we have the center $(a_1 - a_2) = -0.12\,\mu\text{m}$, $(b_1 - b_2) = -1.12\,\mu\text{m}$, and the radius $r = 11.12\,\mu\text{m}$.

Example 20.2. Solve the above problem using the optimization toolbox in MATLAB.

MATLAB provides a function `linprog` that performs LP optimization. It can be used as shown here. We do not have to express our center parameters as differences of two non-negative numbers; that can be accommodated as arguments $[-\text{Inf}, -\text{Inf}, 0]$ and $[\text{Inf}, \text{Inf}, \text{Inf}]$ to the `linprog` function. Pairs $(-\text{Inf}, \text{Inf})$, $(-\text{Inf}, \text{Inf})$, and $(0, \text{Inf})$ correspond to a, b, and r respectively. Therefore A is only of size $n \times 3$, not $n \times 5$. Note that the objective function is defined for a minimization in MATLAB, while the standard LP is a maximization problem. Therefore, F is defined as $[0; 0; -1]$, with the -1 being the coefficient of the radius r. The final argument $[0\,0\,0]$ is the starting solution for the simplex.

```
theta = (0:7)'/8*2*pi;
A = [cos(theta) sin(theta) ones(8,1)];
F = [0;0;-1];
R = [1.1;1.4;1;1.6;1.3;1.2;1.5;1.4]*10;
T=linprog(F,A,R,[],[],[-Inf, -Inf, 0],[Inf Inf Inf],[0 0 0]);
U = T(1)*cos(theta) + T(2)*sin(theta) + T(3);
theta1 = [theta;0];
U1 = [U;U(1)];
polar(theta,R,'rd')
hold on
polar(theta1,U1,'b')
T
```

Table 20.4 Tableau after three iterations of the simplex (final tableau)

	a_1	a_2	b_1	b_2	r	$s(1)$	$s(2)$	$s(3)$	$s(4)$	$s(5)$	$s(6)$	$s(7)$	$s(8)$	R	R/b_2
1	0.00	0.00	−1.00	1.00	0.00	0.29	0.00	−0.71	0.00	0.00	0.41	0.00	0.00	1.12	0.00
2	0.00	0.00	0.00	0.00	0.00	−0.59	1.00	−0.59	0.00	0.00	0.17	0.00	0.00	3.76	0.00
3	0.00	0.00	0.00	0.00	1.00	0.29	0.00	0.29	0.00	0.00	0.41	0.00	0.00	11.12	0.00
4	0.00	0.00	0.00	0.00	0.00	0.41	0.00	−1.00	1.00	0.00	−0.41	0.00	0.00	5.59	0.00
5	0.00	0.00	0.00	0.00	0.00	0.41	0.00	−0.59	0.00	1.00	−0.83	0.00	0.00	1.76	0.00
6	−1.00	1.00	0.00	0.00	0.00	−0.71	0.00	0.29	0.00	0.00	0.41	0.00	0.00	0.12	−1.00
7	0.00	0.00	0.00	0.00	0.00	−0.59	0.00	0.41	0.00	0.00	−0.83	1.00	0.00	2.76	0.00
8	0.00	0.00	0.00	0.00	0.00	−1.00	0.00	0.41	0.00	0.00	−0.41	0.00	1.00	2.17	0.00
E	0.00	0.00	0.00	0.00	0.00	−0.29	0.00	−0.29	0.00	0.00	−0.41	0.00	0.00	−11.12	0.00

20.4 Formulating the MC Problem

The MC circle is the circle of smallest radius that can be constructed so that all measured data points reside on or inside the circle. Mathematically, the problem can be formulated as:

$$\begin{aligned} &\text{Minimize} \quad E = r \\ &\text{such that:} \quad a\cos\theta(i) + b\sin\theta(i) + r \geqslant R(i)\,, \quad \text{for} \quad i = 1:n\,. \end{aligned} \tag{20.5}$$

Notice that because each of the constraints is of greater-than-or-equal-to type, the origin is not a feasible solution. Also notice that the objective function requirement is to minimize, not maximize. A simple trick, again one commonly used in LP, involves multiplying each of the constraints and the objective function by -1 to translate the current formulation into a standard LP (we have also expressed the center parameters a, b as the difference of two non-negative parameters):

$$\begin{aligned} &\text{Maximize}\, E = -r \\ &\text{such that:} \, -(a_1 - a_2)\cos\theta(i) - (b_1 - b_2)\sin\theta(i) - r + s(i) = -R(i)\,, \text{ for } i = 1:n \\ &a_1\,,\, a_2\,,\, b_1\,,\, b_2\,,\, s(i)\,,\, r \geqslant 0\,, \text{ for} \quad i = 1:n\,. \end{aligned} \tag{20.6}$$

The physical interpretation of this sign reversal is that the MC circle of a given set of radial deviations $R(i)$ is in fact the MI circle of $-R(i)$.

There is another approach to solving MC circle problems, one that involves the concept of duality. It offers a link to exchange algorithms and we will discuss that after Example 20.3.

Example 20.3. Find the MC circle center and radius for the data in Example 20.1.

We reverse the sign of the data and find the MI circle as shown below. Notice that while our formulation involves a maximization of $E = -r$, MATLAB's `linprog`

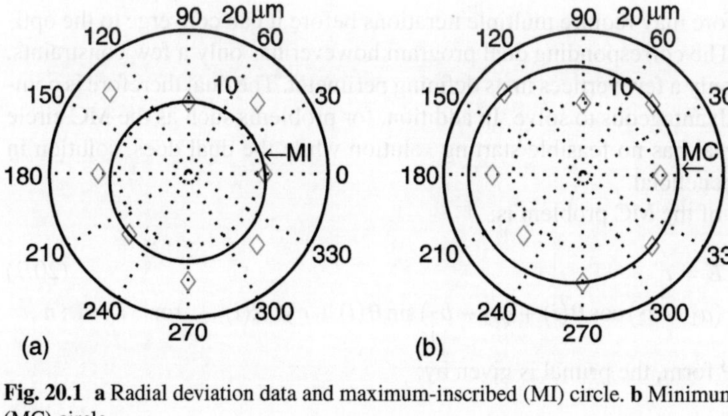

Fig. 20.1 **a** Radial deviation data and maximum-inscribed (MI) circle. **b** Minimum-circumscribed (MC) circle

function expects a minimization problem, therefore $F = [0\ 0\ 1]$ and not $[0\ 0\ -1]$. The MI and MC circles for the data in Example 20.1 are plotted in Fig. 20.1.

```
theta = (0:7)'/8*2*pi;
A = [-cos(theta) -sin(theta) -ones(8,1)];
F = [0;0;1];
R = [1.1;1.4;1;1.6;1.3;1.2;1.5;1.4]*10;
R1 = R*-1;
T=linprog(F,A,R1,[],[],[-Inf, -Inf, 0],[Inf Inf Inf],[0 0 0]);
theta1 = (0:359)'/360*2*pi;
U = T(1)*cos(theta1) + T(2)*sin(theta1) + T(3) ;
theta1 = [theta1;0];
U1 = [U;U(1)];
polar(theta,R,'rd')
hold on
polar(theta1,U1,'b')
T
```

20.5 Duality

Every LP problem, referred to as a primal, has a dual associated with it. As an economic interpretation for a manufacturer of products, the primal is a maximization of sales for the manufacturer given constraints on available resources while the dual is a minimization of cost in acquiring resources. Duality is a very important concept in LP. We urge the reader to review any standard LP textbook. We only describe duality in the context of the MC circle problem here.

For every primal constraint, there is a corresponding dual variable, and for every primal variable, there is a corresponding dual constraint. If the primal has m variables and n constraints, the dual has n variables and m constraints. Because both the primal and the dual converge to the same optimal solution, either can be solved (if it is possible) to determine the optimal values. A primal that features a few variables but a large number of constraints has several vertices in its defining perimeter. The simplex therefore may require multiple iterations before it can converge to the optimal solution. The corresponding dual program however has only a few constraints, and therefore only a few vertices in its defining perimeter. The dual therefore is computationally advantageous to solve. In addition, for problems such as the MC circle where the primal has no feasible starting solution while the dual does, solution in dual space is beneficial.

The primal of the MC problem is:

$$\begin{aligned} &\text{Minimize} \quad E = r \qquad (20.7) \\ &\text{such that:} \quad (a_1 - a_2)\cos\theta(i) + (b_1 - b_2)\sin\theta(i) + r \geqslant R(i)\,, \quad \text{for} \quad i = 1:n\,. \end{aligned}$$

In standard LP form, the primal is given by:

$$\text{Maximize} \quad E = -r \tag{20.8}$$
$$\text{such that:} \quad -(a_1 - a_2)\cos\theta(i) - (b_1 - b_2)\sin\theta(i) - r \leqslant -R(i)\,, \quad \text{for } i = 1:n\,.$$

Its corresponding dual has five constraints and n unknown variables $u(i)$. It is formulated as:

$$\begin{aligned}
\text{Minimize} \quad & U = -u(i)R(i) \\
\text{such that:} \quad & -\sum_{i=1}^{n} u(i)\cos\theta(i) \geqslant 0\,, \\
& \sum_{i=1}^{n} u(i)\cos\theta(i) \geqslant 0 \\
& -\sum_{i=1}^{n} u(i)\sin\theta(i) \geqslant 0\,, \\
& \sum_{i=1}^{n} u(i)\sin\theta(i) \geqslant 0\,, \\
& -\sum_{i=1}^{n} u(i) \geqslant -1
\end{aligned} \tag{20.9}$$

The first two constraints effectively imply one equality constraint. The third and fourth also together imply one equality constraint. Also, by removing the negative signs in the constraints and objective function, we have the following standard LP problem:

$$\begin{aligned}
\text{Maximize} \quad & U = u(i)R(i) \\
\text{such that:} \quad & \sum_{i=1}^{n} u(i)\cos\theta(i) = 0 \\
& \sum_{i=1}^{n} u(i)\sin\theta(i) = 0 \\
& \sum_{i=1}^{n} u(i) \leqslant 1
\end{aligned} \tag{20.10}$$

The dual has a feasible solution at the origin, $u(i) = 0$, for all i. It is also a standard LP program with all constraints of less-than-or-equal-to type and whose objective function is a maximization. The dual can be solved as described in Example 20.1. In general, however, it is solved using the primal table, but in an inverse manner. Whereas in the primal, we first determined the column that would produce the greatest increase in objective function, and then identified the row that limits that increase, here we first identify the limiting row, followed by the column that produces the greatest increase in the objective function. We will not concern ourselves with the details of the solution for the dual; rather we will focus on duality and its implications for exchange algorithms.

Table 20.5 Primal and dual variables and constraints

	$a = (a_1 - a_2)$	$b = (b_1 - b_2)$	r	R	s(i)
$u(1)$	$\cos\theta(1)$	$\sin\theta(1)$	1	$R(1)$	$s(1)$
$u(2)$	$\cos\theta(2)$	$\sin\theta(2)$	1	$R(2)$	$s(2)$
...	...	...	...	...	
$u(n)$	$\cos\theta(n)$	$\sin\theta(n)$	1	$R(n)$	$s(n)$
	0	0	1	0	E
	$v(1)$	$v(2)$	$v(3)$	U	

Table 20.5 shows the primal and dual variables, the primal and dual objective function along with the coefficients of the constraints. The primal slack variables are shown to the right and the dual slack variables are shown in the bottom row. Note that because constraints 1 and 2 in the dual are equality constraints, $v(1) = v(2) = 0$.

The ith primal constraint would be read from Table 20.5 as:

$$a\cos\theta(i) + b\sin\theta(i) + r - R(i) = -s(i)\,. \tag{20.11}$$

The reason for expressing our table in this form is to highlight the relationship between primal and dual variables; variables in a column or in a row are related to each other. That is, primal variable a is tied to dual slack $v(1)$, b to $v(2)$, and r to $v(3)$. Also, dual variables $u(i)$ are related to the corresponding primal slack variables $s(i)$. The nature of the relationship between the variables is brought out by the complementary slackness condition. The theorem, which we simply state here, says at optimality the sum of the products of the pairs of related primal and dual variables is zero. That is,

$$a.v(1) + b.v(2) + r.v(3) + s(1).u(1) + s(2).u(2) + \ldots + s(n).u(n) = 0\,. \tag{20.12}$$

Next, we will demonstrate how this theorem can be used to derive an exchange algorithm for MI and MC circles.

20.6 LP and Exchange Algorithms

As mentioned before, the primal and dual LP reach the same optimal solution. When optimality is reached, the circumscribed (or the inscribed circle) passes through three points (in special cases, the MI or MC circle may contact two points or more than the three). Let the three points have indices $i = p$, q, and r. Then, the following is true for the primal system:

$$\begin{aligned} a\cos\theta(p) + b\sin\theta(p) + r &= R(p) \\ a\cos\theta(q) + b\sin\theta(q) + r &= R(q) \\ a\cos\theta(r) + b\sin\theta(r) + r &= R(r)\,. \end{aligned} \tag{20.13}$$

The slack variables $s(p)$, $s(q)$, and $s(r)$ are zero. The circle does not pass through the other points, and therefore those slack variables are non-zero. From the complementary slackness theorem, this implies that $u(p)$, $u(q)$ and $u(r)$ are non-zero, while $u(i)$ for all values i except p, q, and r, are zero. Because a, b, and r are non-zero, their corresponding dual variables $v(1)$, $v(2)$, and $v(3)$ are zero. Therefore, the corresponding equations in the dual are:

$$\begin{aligned} &u(p)\cos\theta(p)+u(q)\cos\theta(q)+u(r)\cos\theta(r)=0\\ &u(p)\sin\theta(p)+u(q)\sin\theta(q)+u(r)\sin\theta(r)=0\\ &u(p)+u(q)+u(r)=1\,. \end{aligned} \tag{20.14}$$

The solution for this system of equations is:

$$u(p)=\frac{\begin{vmatrix}0 & \cos\theta(q) & \cos\theta(r)\\ 0 & \sin\theta(q) & \sin\theta(r)\\ 1 & 1 & 1\end{vmatrix}}{\begin{vmatrix}\cos\theta(p) & \cos\theta(q) & \cos\theta(r)\\ \sin\theta(p) & \sin\theta(q) & \sin\theta(r)\\ 1 & 1 & 1\end{vmatrix}} =\frac{\sin(\theta(r)-\theta(q))}{\sin(\theta(r)-\theta(q))+\sin(\theta(q)-\theta(p))+\sin(\theta(p)-\theta(r))}$$

$$u(q)=\frac{\begin{vmatrix}\cos\theta(p) & 0 & \cos\theta(r)\\ \cos\theta(p) & 0 & \sin\theta(r)\\ 1 & 1 & 1\end{vmatrix}}{\begin{vmatrix}\cos\theta(p) & \cos\theta(q) & \cos\theta(r)\\ \sin\theta(p) & \sin\theta(q) & \sin\theta(r)\\ 1 & 1 & 1\end{vmatrix}} =\frac{\sin(\theta(p)-\theta(r))}{\sin(\theta(r)-\theta(q))+\sin(\theta(q)-\theta(p))+\sin(\theta(p)-\theta(r))}$$

$$u(r)=\frac{\begin{vmatrix}\cos\theta(p) & \cos\theta(q) & 0\\ \cos\theta(p) & \sin\theta(q) & 0\\ 1 & 1 & 1\end{vmatrix}}{\begin{vmatrix}\cos\theta(p) & \cos\theta(q) & \cos\theta(r)\\ \sin\theta(p) & \sin\theta(q) & \sin\theta(r)\\ 1 & 1 & 1\end{vmatrix}} =\frac{\sin(\theta(q)-\theta(p))}{\sin(\theta(r)-\theta(q))+\sin(\theta(q)-\theta(p))+\sin(\theta(p)-\theta(r))} \tag{20.15}$$

To be valid, $u(p)$, $u(q)$, and $u(r)$ should satisfy the non-negativity constraint. It can be shown that this will be satisfied when $\sin(\theta(r)-\theta(q))>0$, $\sin(\theta(q)-\theta(p))>0$, $\sin(\theta(p)-\theta(r))>0$, or the difference between consecutive angles be less than or equal to 180°. Under this condition, the numerator is positive and all three terms of the denominator are also positive; therefore the ratio is positive. In summary, when optimality is reached, the three points on MI or MC circle are arranged so that no two points subtend an angle greater than 180° at the center.

20.7 Exchange Algorithms for MI and MC Circles

The derivation above suggests a simple exchange algorithm that can be implemented for inscribed and circumscribed circles. Three random points are chosen to initiate the algorithm. Then, the outlier point that lies farthest away is determined. One of the three initial points is replaced with the new point so the 180 rule is not violated. We leave it an exercise to the reader to implement the 180 rule.

20.8 Summary

The determination of the MZ, MC, and MI boundaries are essentially optimization problems. In roundness metrology, we encounter linear approximations for the circle and the cylinder, and we can therefore conveniently employ LP optimization techniques such as the simplex algorithm to find the solutions. In this chapter, we have reviewed how LP simplex can be used to determine the MI and MC circle center and radius. A similar approach can be used for MZ circles as well, and for the cylinder geometry. We also showed in this chapter how the LP formulation is linked to exchange algorithms, and derived an algorithm for MC and MI circles.

Exercises

20.1. Determine the parameters of the MC circle for the data in Example 20.1 by solving the dual.

20.2. Write a program to implement the 180 rule.

20.3. Write a program to fit a limaçon through any three given points.

20.4. Using the exchange algorithm programs written in Exercises 20.2 and 20.3 determine the MI and MC circle centers and radii for the data in Example 20.1.

20.5. The MZ circle can be formulated as

$$\begin{aligned} &\text{Minimize} \quad E = h \\ &\text{such that:} \quad a\cos\theta(i) + b\sin\theta(i) + r + h \geqslant R(i)\,, \quad \text{for} \quad i = 1:n \\ &\qquad\qquad\quad a\cos\theta(i) + b\sin\theta(i) + r - h \leqslant R(i)\,, \quad \text{for} \quad i = 1:n \end{aligned}$$

Express this as a standard LP problem, with the center parameters as a difference of two non-negative parameters. Is the origin $(a,b,r,h) = (0,0,0,0)$ a feasible solution? State the dual for this problem and show that the origin in the dual is a feasible solution.

20.6. Determine the MZ center, radius, and zone width $2h$ for the data in Example 20.1 using MATLAB's `linprog` function.

20.7. Verify the value in Exercise 20.5 is consistent with that obtained using the exchange algorithm described in Chap. 19.

20.8. What are the indices of the points that contact the zone for the data in Example 20.1? How can you determine this from the output of the `linprog` function in Exercise 20.6?

20.9. For the data below, using the linear approximation for the limaçoid given in Chap. 18 (Eq. 18.10):

Angle (rad)	0	0.7854	1.5708	2.3562	3.1416	3.9270	4.7124	5.4978
Radius (μm)	11	14	10	16	13	12	15	14
z (mm)	0	0	0	0	0	0	0	0

Angle (rad)	0	0.7854	1.5708	2.3562	3.1416	3.9270	4.7124	5.4978
Radius (μm)	12	11	13	16	10	12	12	11
z (mm)	1	1	1	1	1	1	1	1

Angle (rad)	0	0.7854	1.5708	2.3562	3.1416	3.9270	4.7124	5.4978
Radius (μm)	16	12	10	11	14	17	10	11
z (mm)	2	2	2	2	2	2	2	2

(a) Formulate the MI cylinder in its primal form and determine the center, tilt, and radius
(b) Formulate the MC cylinder in its dual form and determine the parameters
(c) Formulate the MZ cylinder in its dual form and determine the zone width
(d) Suggest a procedure to determine the center, tilt, and radius of the MZ cylinder from the solution of the dual

References

Chetwynd, D.G. 1985, 'Applications of linear programming to engineering metrology', *Proceedings of the Institution of Mechanical Engineers Part B: Management and Engineering Manufacture*, vol. 199, no. B2, pp. 93–100.

Ignizio, J.P. 1981, *Linear Programming in Single- & Multiple-Objective Systems*, Prentice-Hall. Englewood Cliffs, NJ, USA.

Sultan, A. 1993, *Linear Programming: An Introduction with Applications*, Academic Press. San Diego, CA, USA.

Part IV
Parameterization

Parameters extracted from surface profiles and areal topographies constitute actual quantitative information that may be used as the link between a component's functionality, the manufacturing process variation, and surface texture. Therefore, a variety of parameters have been defined over the years, some very specific to certain applications, others more generally used in a number of situations. In Part IV, we describe computational aspects of determining parameters for 2D profiles and 3D topographies. We discuss amplitude, spacing and hybrid parameters from 2D profiles in Chap. 21 and statistical function based parameters in Chap. 22. We extrapolate these for 3D topographies in Chap. 23.

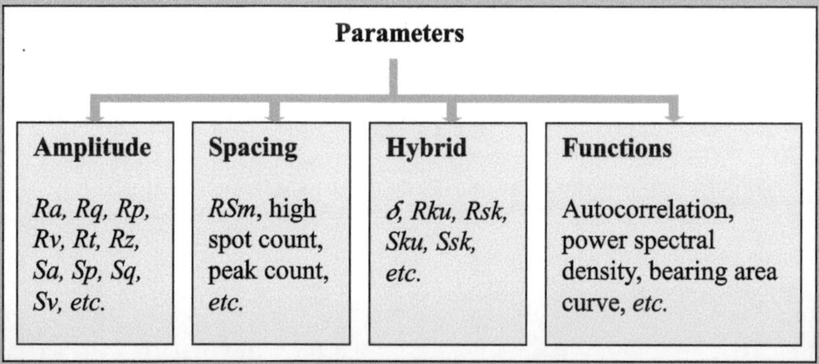

Fig. IV Some 2D and 3D parameters which we will compute in Part IV

Chapter 21
Surface Finish Parameters I: Amplitude, Spacing, Hybrid, and Shape

21.1 Introduction

The purpose of defining parameters is to succinctly extract meaningful information from surface profiles that could then be related to the functional performance of the part. The information from the profile could describe amplitude characteristics, spacing characteristics or some characteristics associated with the general shape of the profile. Over the years, there has been a profusion of parameters defined by industry for specific needs (Whitehouse 1982), and this has resulted in considerable confusion as to which parameter is suitable for any given application. It is not the purpose of this chapter to present a compendium of parameters and definitions; rather, our objective is to describe computational aspects of some commonly used parameters. On the basis of the information provided here, the reader can hopefully implement parameters of his or her choosing using a variety of mathematics packages or programming languages. The parameters we describe are from the American (American Society of Mechanical Engineers 2002) and the international (International Organization for Standardization 1997) Standards.

21.2 Amplitude Parameters

Let r represent a roughness profile containing n elements and $r(i)$ be the ith element of r. Then, below is a list of some amplitude parameters (American Society of Mechanical Engineers 2002):

1. Average roughness, *Ra,* is defined as the arithmetic average of the absolute values of the roughness profile r, measured from the mean line. Thus,

$$Ra = \frac{1}{n}\sum_{i=1}^{n} |r(i)| \,. \tag{21.1}$$

2. Root mean square roughness, *Rq*, is defined as the root mean square average of the profile *r* measured from the mean line. Thus,

$$Rq = \sqrt{\frac{1}{n}\sum_{i=1}^{n} r(i)^2}\,. \tag{21.2}$$

3. Maximum profile peak height, *Rp*, is defined as the distance to the highest point of the profile measured from the mean line, see Fig. 21.1.
4. Maximum profile valley height, *Rv*, is defined as the distance to the lowest point of the profile measured from the mean line, see Fig. 21.1.
5. Maximum height of the profile, *Rt*, is defined as the distance between the highest and lowest points of the profile.
6. Average maximum height of the profile, *Rz*, is defined as the average of the successive values of Rt_i, calculated over the length of the profile. Rt_i is defined as the maximum height of the profile within the *i*th sampling length. Sampling length is equal to the cutoff for parameters computed on roughness and waviness profiles. For parameters computed on a raw profile, the evaluation length is used instead of sampling length. Therefore *Rz* is simply equal to *Rt*. Figure 21.1 illustrates this.

It should be noted that the computational challenges in determining many commonly used parameters lie not in their mathematical complexity, but in understanding and conforming to the definition given in any Standard. For instance, the American and international Standards have different ways of determining even simple parameters such as *Ra*; the ASME B46.1 (American Society of Mechanical Engineers 2002) defines *Ra* as computed over the evaluation length, which is the

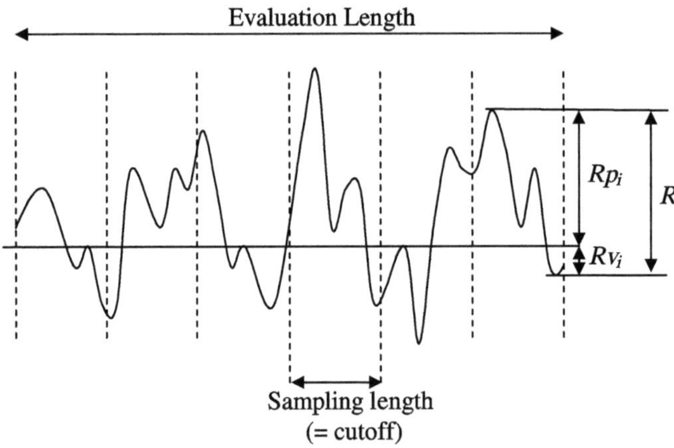

Fig. 21.1 Evaluation length, sampling length, and some amplitude parameters. Rt_i is the maximum height of the profile within the *i*th sampling length, Rp_i is the distance to the highest point of the profile measured from the mean line, Rv_i is the distance to the lowest point of the profile measured from the mean line

measured profile length while the ISO 4287 (International Organization for Standardization 1997) defines the parameter as computed over the sampling length, which is typically equal to a cutoff. *Rz* has multiple definitions in different Standards (Tabenkin 2007).

Example 21.1. Write a MATLAB function that accepts a profile *r* and returns the amplitude parameters *Ra*, *Rq*, *Rp*, *Rv*, and *Rt* in that order, computed on profile *r*, and evaluated over the evaluation length.

```
function [Ra, Rq, Rp, Rv, Rt] = myAmplitudeParameters(r)
n = size(r,1);              % determine length of profile r
r = r - mean(r);            % shift profile to zero mean
Ra = sum(abs(r))/n;         % compute Ra
Rq = sqrt(sum(r.*r)/n);     % compute Rq
Rp = max(r);                % compute Rp
Rv = abs(min(r));           % compute Rv
Rt = Rp + abs(Rv);          % compute Rt
```

Example 21.2. Generate profile z_1 of amplitude 1 μm and wavelength 2 mm, over a length of 8 mm with a sampling interval $\Delta x = 1$ μm. Generate profile z_2 of amplitude 1 μm and wavelength 0.5 mm, with a sampling interval $\Delta x = 1$ μm. Do the amplitude parameters distinguish the two profiles?

```
% Generate profile z1 and z2
dx = 0.001;                % spacing in mm
l = 8;                     % length in mm
n = l/dx;                  % number of profile points
x = (0:dx:l -dx)';
z1 = 1*sin(2*pi.*x/2);     % generate z1
z2 = 1*sin(2*pi.*x/0.5);   % generate z2
% Compute amplitude parameters
[Ra1, Rq1, Rp1, Rv1, Rt1] = myAmplitudeParameters(z1);
[Ra2, Rq2, Rp2, Rv2, Rt2] = myAmplitudeParameters(z2);
```

The results are shown below

```
[Ra1, Rq1, Rp1, Rv1, Rt1]
0.6366 0.7071 1.0000 1.0000 2.0000
[Ra2, Rq2, Rp2, Rv2, Rt2]
0.6366 0.7071 1.0000 1.0000 2.0000
```

It is clear that the amplitude parameters are identical for both profiles and therefore do not distinguish the two profiles. There is a need for parameters that capture spacing information, and these are discussed in Sect. 21.3.

21.3 Spacing Parameters

While amplitude parameters provide a measure of the height or depth of the profile elements from the mean line, spacing parameters provide a measure of the spacing between the peaks and valleys. There are several spacing parameters; computational aspects of a few of the commonly used are described here:

1. *RSm* is the mean peak spacing (American Society of Mechanical Engineers 2002), and is defined as the average distance between zero crossings of the profile. If there are q zero crossings, then

$$RSm = \frac{1}{q}\sum_{i=1}^{q} Sm(i)\,. \tag{21.3}$$

   Note, however, that a zero crossing shall only be considered if two conditions are met: (a) the maximum profile height between the two zero crossings is greater than a threshold (typical value is 10% of *Rz*), and (b) the spacing between the two zero crossings is greater than a threshold (typical value is 1% of the sampling length). Figure 21.2 illustrates the computation of *RSm*. It should be noted that there are discrepancies in the different definitions of *RSm* (Leach and Harris 2002).
2. High spot count (International Organization for Standardization 1997) is the ratio of the number of peaks that cross a certain threshold to the evaluation length. A peak is considered if the profile rises above a threshold and falls back below the same threshold. In Fig. 21.2, the vertical arrows show high spots. There are seven high spots that cross the upper threshold.
3. Peak count (International Organization for Standardization 1997) is similar to high spot count, but also involves a lower threshold. Thus, the profile must rise above the upper threshold and fall below the lower threshold to be counted. For implementation purposes, we determine all peaks that cross the upper threshold. Then, we examine pairs of the peaks to find out if there is any point between them where the profile falls below the lower threshold. If so, that peak is counted, otherwise it is skipped. In Fig. 21.2, there are seven high spots that cross the upper threshold. Of these, high spots 3 and 5 cannot be considered for peak count purposes because the intermediate valleys (between 2 and 3, and between 4 and 5) do not fall below the lower threshold.

Example 21.3. Write a MATLAB function to determine *RSm*. The function should accept four arguments: the profile array r, spacing parameter Δx, the height, and spacing thresholds in micrometers.

In order to compute *RSm*, we have to first identify pairs of zero crossings that satisfy the following conditions:

1. There is a peak between them that rises above the upper threshold.
2. There is a valley between them that falls below the lower threshold.
3. The distance between the zero crossings is greater than the spacing threshold.
4. There is also one other condition that is not explicitly stated but implied: between the second zero crossing and the one immediately following that, there must be a peak that crosses the upper threshold.

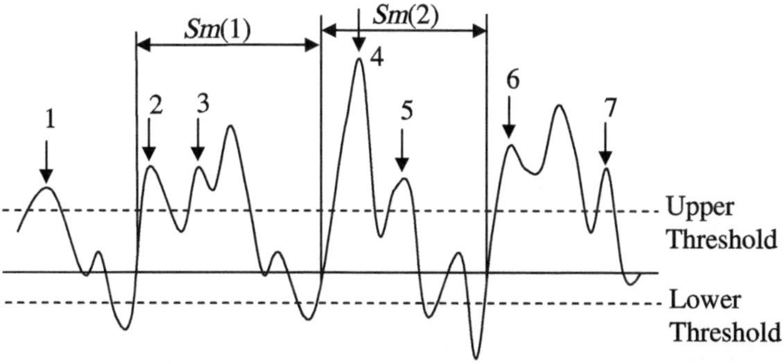

Fig. 21.2 Spacing parameter *Sm* is determined by averaging *Sm*(1), *Sm*(2), *etc*. The *vertical arrows* show high spots

This is illustrated in Fig. 21.3. There are five zero crossings for the profile shown. We identify the first zero crossing with positive slope (point 1 in Fig. 21.3). We set this zero crossing as the reference for determining $Sm(1)$. Point 2 is discarded as the right boundary for $Sm(1)$, because it violates rule 2 above. Point pairs 1 and 3 satisfy rules 1 and 2 but fail to satisfy rule 4, because there is no peak between points 3 and 4 that rise above the upper threshold. Therefore point 3 is also discarded. For the same reason, point 4 is also not considered. Finally, we consider point pairs 1 and 5, which satisfy rules 1, 2, and 4. If the distance between points 1 and 5 is greater than the spacing threshold, we consider that distance as $Sm(1)$. As a matter of detail, we do not directly consider the distance between points 1 and 5 as $Sm(1)$, as they do not necessarily lie on the zero line. Instead, we interpolate to determine the axial location where the profile intersects the x axis and consider the distance between the interpolated points as $Sm(1)$. This process is now repeated for $Sm(2)$, by setting point 5 as the reference, and a search is initiated for determining the right boundary in the manner described above. The following MATLAB function describes this process.

```
function RSm = findRSm(r, dx, HeightThreshold, SpacingThreshold)
n = size(r,1);
% Find all zero crossing
k = 1;
for j=1:n-1
```

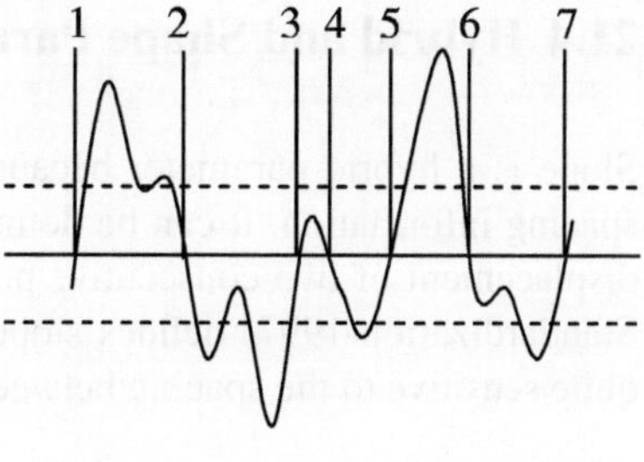

Fig. 21.3 Calculation procedure for the mean peak spacing (*RSm*)

```
    if (r(j,1)*r(j+1,1) < 0)
        B(k,1) = j;
        k = k +1;
     end
end
% Make sure the first zero is a~low to high zero crossing
if r(B(1,1),1) > 0 & r(B(1,1)+1,1) < 0
    B = B(2:size(B,1),1);
end
% Compute RSm
Left = 1;
Right = Left + 2;
NextToRight = Left + 3;
k = 1;
while NextToRight <= size(B,1)
    a = max(r(B(Left,1):B(Right,1),1));
    b = min(r(B(Left,1):B(Right,1),1));
    c = max(r(B(Right,1):B(NextToRight,1),1));
    if a > HeightThreshold & b < -1*HeightThreshold
& c > HeightThreshold
        LeftInterpolated = B(Left,1)*dx - dx*r(B(Left,1))/
(r(B(Left,1)+1)-r(B(Left,1)));
        RightInterpolated = B(Right,1)*dx - dx*r(B(Right,1))/
(r(B(Right,1)+1)-r(B(Right,1)));
        temp = RightInterpolated - LeftInterpolated;
        if temp > SpacingThreshold
            Smi(k,1) = temp;
            k = k + 1;
            Left = Right;
            Right = Left + 2;
            NextToRight = Left + 3;
        else
            Right = Right + 2;
            NextToRight = NextToRight + 2;
        end
    else
        Right = Right + 2;
        NextToRight = NextToRight + 2;
    end
end
RSm = sum(Smi)/size(Smi,1);
```

21.4 Hybrid and Shape Parameters

Slope is a hybrid parameter because it captures a combination of amplitude and spacing information. It can be defined as the ratio of the vertical to the horizontal displacement of two consecutive points. ISO 4287 (International Organization for Standardization 1997) defines slope using six consecutive points as follows. It is quite sensitive to the spacing between profile points.

$$\delta = \tan^{-1}\left(\sqrt{\frac{\sum_{j=4}^{n-3} \Delta(j)^2}{(n-6)}}\right)\frac{180}{\pi}, \tag{21.4}$$

where

$$\Delta(j) = \frac{r(j-3) - 9r(j-2) + 45r(j-1) - 45r(j+1) + 9r(j+2) - r(j+3)}{60\Delta x}. \tag{21.5}$$

Skew is a measure of the asymmetry of the profile and is defined as

$$Rsk = \frac{1}{nRq^3}\sum_{i=1}^{n} r(i)^3, \tag{21.6}$$

and kurtosis is a measure of the spikiness of the profile. It is defined as

$$Rku = \frac{1}{nRq^4}\sum_{i=1}^{n} r(i)^4. \tag{21.7}$$

Example 21.4. Generate the following three profiles: (a) sinusoidal profile z_1 of amplitude 1 μm and wavelength 0.8 mm, over a length of 8 mm at a sampling interval $\Delta x = 1$ μm; (b) profile z_2 generated by taking the absolute value of z_1 and shifting it so that the mean is zero; (c) profile z_3 generated by mirroring the profile z_2 about the x axis. Compute the skew for the three profiles and explain the observations.

The following MATLAB program generates the profiles and computes skew. From the results, it is seen that profile z_1 has zero skew as expected because it is distributed uniformly on both sides of the profile. Profile z_2 is negatively skewed as seen in Fig. 21.4 (the mean line is closer to the top rather than the bottom) and the parameter reflects this. Likewise, z_3 is positively skewed; more of the profile appears to be above the mean line than below.

```
% Computing skew for three profiles
dx = 0.001;       % spacing in mm
l = 8;            % length in mm
n = l/dx;         % number of profile points
x = (0:dx:l-dx)';
% part a)
z1 = 1*sin(2*pi.*x /0.8);            % generate profile z1
Rq1 = sqrt(sum(z1.^2)/n);            % compute Rq
Rsk1 = (1/Rq1^3)*sum(z1.^3)/n;       % compute Rsk
% part b)
z2 = abs(z1);                        % generate profile z2
z2 = z2-mean(z2);                    % shift it to zero mean
Rq2 = sqrt(sum(z2.^2)/n);            % compute Rq
Rsk2 = (1/Rq2^3)*sum(z2.^3)/n;       % compute Rsk
% part c)
z3 = z2*(-1);                        % generate profile z3
Rq3 = sqrt(sum(z3.^2)/n);            % compute Rq
Rsk3 = (1/Rq3^3)*sum(z3.^3)/n;       % compute Rsk
```

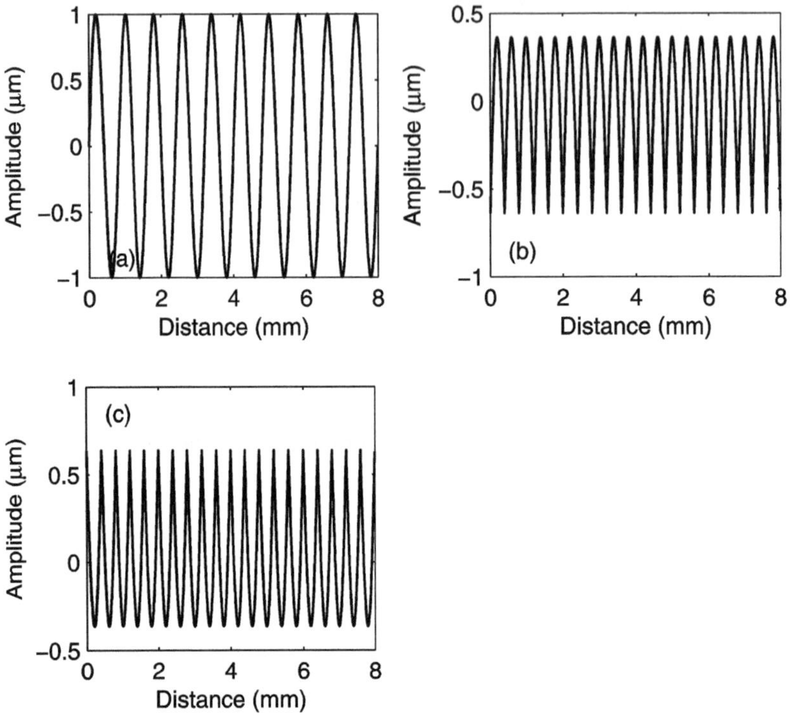

Fig. 21.4a–c Profiles z_1, z_2 and z_3. **a** z_1 has zero skew. **b** z_2 is negatively skewed ($Rsk = -0.4972$). **c** z_3 is positively skewed ($Rsk = 0.4972$)

```
% plot
plot(x,z1,'r')
xlabel('Distance (mm)');
ylabel('Amplitude (\mum)');
plot(x,z2,'r')
xlabel('Distance (mm)');
ylabel('Amplitude (\mum)');
plot(x,z3,'r')
xlabel('Distance (mm)');
ylabel('Amplitude (\mum)');
```

21.5 Summary

We have discussed a variety of parameters that can be computed from surface profiles. Some of these parameters capture amplitude characteristics, others capture spacing information, and some are intended to reflect information about the shape of the profile such as its spikiness, asymmetry, or slope. As a general guideline, amplitude parameters are useful for random profiles and spacing parameters for more periodic profiles. There are also other parameters that can be computed from functions derived from surface profiles. These are discussed in Chap. 22.

Exercises

21.1. Write a MATLAB function that accepts a profile z, the cutoff λc and spacing Δx as input arguments, and outputs the amplitude parameter Rz computed on profile z (do not perform any filtering here). Ensure that if λc is a null value or zero, Rz is computed over the evaluation length (and therefore simply equal to Rt).

21.2. Filter profile z_1 in Example 21.2 using a Gaussian filter with three different cutoffs – 0.25 mm, 0.8 mm, and 2.5 mm. Verify that the amplitude parameters (Ra and Rq) increase as cutoff is increased and that the waviness parameters (Wa and Wq) decrease as cutoff is increased. Simplify the evaluation process by writing a function that performs the filtering and parameter computations. Note that parameters are generally computed after discarding a length equal to one cutoff from the beginning and end of both the roughness and waviness profiles (filtering edge effects).

21.3. For a sinusoidal profile of wavelength λ (assume $l = 4\lambda$), will RSm computed on the raw profile be identically equal to λ? Verify your answer in MATLAB for a profile z of amplitude 1 μm and $\lambda = 2$ mm ($\Delta x = 1$ μm, $n = 8{,}000$ points). Assume the spacing threshold is 0.5 mm and the amplitude threshold is 0.2 μm. Introduce a small phase offset in the data (say $\lambda/8$) and re-compute RSm. Will RSm always be equal to 2 mm regardless of phase offset? Explain.

21.4. Write a MATLAB function to compute the slope parameter as defined in Sect. 21.4. Compute the slope parameter for two sinusoidal profiles z_1 (amplitude 2 μm, wavelength 0.4 mm) and z_2 (amplitude 0.5 μm, wavelength 2 mm) generated over a length of 8 mm at a spacing of 1 μm. Explain the results.

21.5. Generate three profiles as described here: (a) a random profile z_1 (use MATLAB function `normrnd` in statistical toolbox, zero mean, and 0.3 μm standard deviation) containing 8,000 points. Let $\Delta x = 1$ μm. Shift the profile so that the mean is zero; (b) a triangular profile z_2 of period 1 mm as shown below:

```
x = (0:0.001:7.999)';
D = 0:1:8;                        % repetition frequency
z2 = pulstran(x,D,'tripuls'); % generate triangular pulse train
z2 = z2 - mean(z2); plot(z2)
```

(c) A profile z_3 determined as the fifth power of z_2. Compute kurtosis on the three profiles and explain the results.

References

American Society of Mechanical Engineers 2002, *ASME B46.1-2002, Surface Texture (Surface Roughness, Waviness and Lay)*, ASME. New York, NY, USA.

International Organization for Standardization 1997, *ISO 4287:1997, Geometrical Product Specifications (GPS) – Surface texture: Profile Method – Terms, Definitions and Surface Texture Parameters*, ISO. Geneva, Switzerland.

Leach, R.K. and Harris, P.M. 2002, 'Ambiguities in the definition of spacing parameters for surface-texture characterization', *Measurement Science and Technology*, vol. 13, pp. 1924–1930.

Tabenkin, A. 2007, 'The ABCs of Rz', Quality Magazine, October 29th.

Whitehouse, D.J. 1982, 'The parameter rash – is there a cure?', *Wear*, vol. 83, no. 1, pp. 75–78.

Chapter 22
Surface Finish Parameters II: Autocorrelation, Power Spectral Density, Bearing Area

22.1 Autocovariance and Autocorrelation Function

In order to determine how repetitive a profile is, a profile can be compared against itself, but shifted in phase by small amounts. A function that tracks the correlation of a profile against itself is termed the autocovariance function (*ACV*) [see ASME B46.1 (American Society of Mechanical Engineers 2002)]. A copy of a profile is first overlapped with the original and a coefficient is computed as the sum of the term-by-term product of profile elements in the original and duplicate profiles. Then, the copy is shifted by some amount and a new coefficient is evaluated. These coefficients, plotted as a function of shift distance, form the *ACV*.

Formally, *ACV* is defined as the integral of the shifted and unshifted profiles evaluated over the length of the profile. For discrete profile data, *ACV* is defined as

$$ACV(k) = \frac{1}{n}\sum_{j=1}^{n-k} z(j)z(j+k)\,, \quad 0 \leqslant k \leqslant n-1\,. \tag{22.1}$$

As the copy of the original profile is shifted, the overlap area decreases. Equation 22.1 does not account for this reducing overlap area and is therefore termed the "biased" autocovariance. The "unbiased" autocovariance that accounts for this reducing overlap region is given by

$$ACV(k) = \frac{1}{n-k}\sum_{j=1}^{n-k} z(j)z(j+k)\,, \quad 0 \leqslant k \leqslant n-1\,. \tag{22.2}$$

The resulting *ACV* contains *n* elements and the *ACV* plot is sometimes referred to as the "right-side" plot because the duplicate profile is shifted from the maximum overlap region (zero shift) along one direction only until there is no overlap. The duplicate profile when shifted in the opposite direction from the maximum overlap region produces the "left-side" *ACV*. It can be generated from the following:

$$ACV(k) = \frac{1}{n+k}\sum_{j=1-k}^{n} z(j)z(j+k)\,, \quad -n+1 \leqslant k \leqslant -1\,. \tag{22.3}$$

A combination of Eqs. 22.2 and 22.3 can be used to generate the complete *ACV* for any given profile. Unless otherwise stated, we will use the "biased right-side" estimate for the autocovariance (Eq. 22.1). The autocorrelation function (*ACF*) is the *ACV* normalized by the square of the root mean square roughness. Therefore,

$$ACF(k) = \frac{ACV(k)}{Rq^2} . \tag{22.4}$$

The correlation length is the shift distance at which the *ACF* falls to a set threshold. Typical threshold values are 1/e (natural logarithm), 0.1 or zero. The use of the *ACF* for describing and classifying surfaces was first reported by Peklenik (1967). In practice, profiles are overlapped by only about 10% of their lengths; any further overlap results in fast decay of the *ACV*.

Example 22.1. Write a MATLAB function to generate the *ACV* and *ACF* given by Eqs. 22.1 and 22.4.

The statistical toolbox in MATLAB provides an inbuilt function `xcorr` that can be used to compute *ACF*. It is also fairly straightforward to develop a function that computes the *ACV* and *ACF* as shown below:

```
function [ACV, ACF, shift] = computeACF(z,dx)
n = size(z,1);            % length of profile z
Rq = sqrt(sum(z.*z)/n); % determine Rq
for k = 1:n
    temp = 0;
    for j=1:n-k           % compute the sum of products
        temp = temp + z(j,1)*z(j+k-1,1);
    end
    ACV(k,1) = temp/n;   % divide by number of points
end
shift = (0:n-1)'*dx;      % generate the x axis array of
                          % shift distances
ACF = ACV/Rq^2;
```

Example 22.2. Compute the correlation length for a sinusoidal profile z (amplitude 1 μm, wavelength 2 mm) generated in the interval $0 \leqslant x < 8$ mm at a spacing $\Delta x =$ 1 μm. Assume threshold is zero. Interpret the plot.

Profile z is generated and *ACF* plotted (see Fig. 22.1) as shown below:

```
% ACF
dx = 0.001;              % spacing in mm
l = 8;                   % length in mm
n = l/dx;                % number of profile points
x = (0:dx:l-dx)';
z = 1*sin(2*pi.*x/2); % generate profile z
[ACV, ACF, shift] = computeACF(z, dx);    % compute ACF and ACV
CorrelationLength = min(find(ACF<=0))*dx; % compute correlation
                                          % length
plot(shift,ACF);
xlabel('Shift distance (mm)');
ylabel('ACF');
```

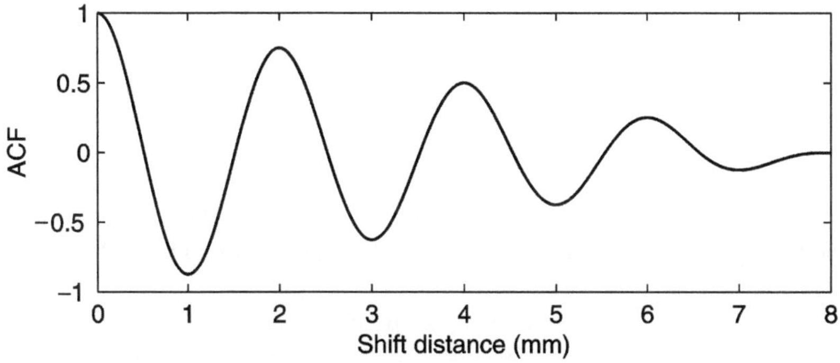

Fig. 22.1 Autocorrelation function (*ACF*) as function of shift distance

The correlation length can be computed as the distance from the origin to where the *ACF* drops below the threshold of zero; the value in this case is 0.515 mm. This value can be refined by interpolating near the zero crossing to obtain the exact location where the *ACF* crosses the x axis. We leave it as an exercise to the reader to perform this.

An interesting aspect of the *ACF* plot is the periodicity evident in Fig. 22.1. As the shift distance is increased from zero, the correlation diminishes until it is out of phase by $\pi/2$, when the original profile and its copy are exactly out of phase and *ACF* drops to zero. As the shift increases even further, the two profiles fall out of phase and again realign in phase, resulting in the periodic pattern seen in Fig. 22.1. Because the input profile has a period of 2 mm, the *ACF* reflects the same period of 2 mm. The amplitude of the *ACF* drops as the shift increases because the two profiles have increasingly smaller overlapping areas. If the two profiles were infinitely long, the amplitude of the *ACF* would not taper off.

As much as this discussion is useful in understanding the *ACF* plot, the *ACF* is typically not used for periodic profiles such as sinusoids. In fact, its greater use lies in capturing any hidden periodicity in more random profiles, such as those produced by grinding where the correlation length has some relationship with the grain structure of the grinding wheel (Whitehouse 1994). For periodic profiles, the Fourier transform or the power spectral density (*PSD*) are more useful tools. The *PSD* is discussed next.

22.2 Power Spectral Density

The *PSD* is the square of the Fourier transform of the profile, appropriately normalized (American Society of Mechanical Engineers 2002). For a discrete profile z whose discrete Fourier transform is given by zf, the *PSD* is given by

$$PSD(f) = \frac{\Delta x}{n} |zf(f)|^2 , \tag{22.5}$$

where f is the spatial frequency (reciprocal of the spatial wavelength) given by k/l, k is an integer ranging from 1 to $n/2$. The area under the *PSD* curve is equal to Rq^2. The *PSD* can also be computed from Fourier transform of the *ACV*. These are illustrated in the following examples.

Example 22.3. Plot the *PSD* of profile z in Example 22.2.

Profile z is generated as shown below. The *PSD* is computed and plotted as a function of both frequency and surface wavelength. These are shown in Fig. 22.2.

```
% PSD
n = 8000;                  % number of profile points
dx = 1;                    % spacing
x = (0:dx:n-1)'*dx;
z = 1*sin(2*pi.*x/2000); % generate profile z
zf = fft(z);               % evaluate DFT of z
PSD = dx/n*(abs(zf).^2);
j = (2:1:floor(n/2)+1)';
freq = (j-1)/(n*dx);       % generate frequency array
semilogx(freq,PSD(2:floor(n/2)+1,1)); % plot PSD as function of
                                      % frequency
xlabel('Frequency (1/\mum)');
ylabel('PSD (\mum^3)');
wave = n*dx./(j-1);                   % generate wavelength array
plot(wave, PSD(2:floor(n/2)+1,1));% PSD as function of wavelength
xlabel('Wavelength (\mum)');
ylabel('PSD (\mum^3)');
```

Example 22.4. Compute the area under the *PSD* and show that it is equal to Rq^2.

The area under the *PSD* curve can be computed numerically from the *PSD* plotted as a function of frequency. The frequency step is $1/l$. Therefore, the sum of elements of the *PSD* array times the frequency step is the area under the curve. For profile z, this value is 0.5. Rq^2 for z is equal to 0.5. Note that the *PSD* array contains 8,000 elements and the area is computed over the entire *PSD* array length, not just the one half spectrum plotted in Fig. 22.2.

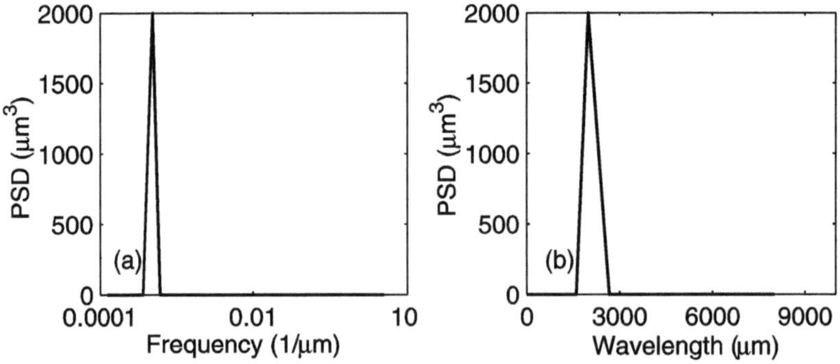

Fig. 22.2a,b Power spectral density (*PSD*) of profile z **a** as a function of frequency and **b** as a function of wavelength

```
Area_PSD = sum(PSD)*1/(n*dx); % evaluate area under PSD (plotted
                              % as function of frequency)
```

Example 22.5. Compute the discrete Fourier transform of the *ACV* for profile *z* and show that it is identical to the *PSD*.

In order to estimate the *PSD* from the *ACV*, we require the "unbiased left- and right-side" estimate for the *ACV* (combination of Eqs. 22.2 and 22.3). While the program written in Example 22.1 can be easily modified, we use MATLAB's `xcorr` function in the statistical toolbox to compute the *ACV*. Notice that the resulting array's length is one less than twice the length of array *z*. This is because the autocovariance array has been computed by shifting the profile from its maximum overlap area (zero shift) along both directions. From this array, we extract the central *n* elements and perform its discrete Fourier transform (DFT) as shown below. The resulting *PSD* array can be seen to be identical to that generated in Example 22.3.

```
C = xcorr(z,'unbiased');
C = C(4001:12000);
PSD1 = abs(fft(C));
```

22.3 Amplitude Density Function and Bearing Area Curve

The amplitude density function (*ADF*) is the histogram of the profile and is computed over the evaluation length, as defined in the ASME B46.1 Standard (American Society of Mechanical Engineers 2002). The vertical range of the profile is partitioned into *q* bins. The number of profile points contained in each bin is plotted as a function of location of center of the bin to obtain the histogram. The bearing area curve (*BAC*) is the cumulative distribution of the *ADF*, and it is also called the Abbott–Firestone curve. Figure 22.3 illustrates the *ADF* and *BAC* for a profile. *Htp* is the difference in heights for two profile bearing length ratios *tp*, as shown in Fig. 22.3. The following examples illustrate the computation of the *ADF* and the *BAC*.

Example 22.6. Write a function to plot the histogram of a profile *z*.

The function accepts two arguments, the profile and the number of bins, as inputs. The output arrays are the frequency count (number of points in each bin *H*) and the location of the bin (center of the bin).

```
% ADF
function [H,binLocation] = ADF(z, numBins)
a = max(z);                   % determine the maximum value in z
b = min(z);                   % determine the minimum value in z
binWidth = (a-b)/numBins;     % determine bin width
t = 1;
for k=b:binWidth:a-binWidth
```

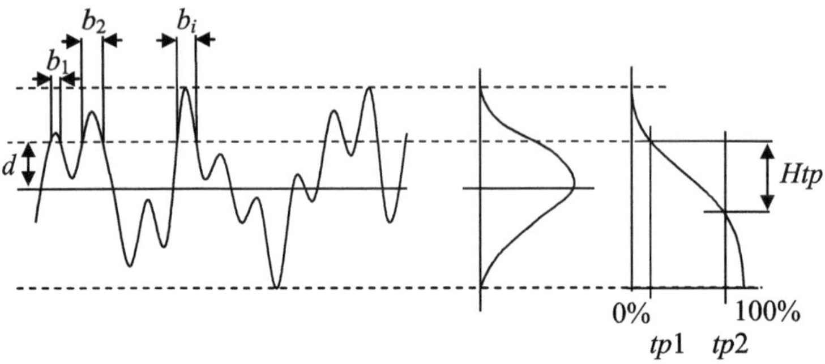

Fig. 22.3 Example profile and its amplitude density function (*ADF*) and bearing area (*BAC*). The sum of the profile lengths $b_1, b_2, \ldots, b_i$, divided by the spacing will give the number of points that lie above the line at a distance d from the mean (cumulative distribution)

```
    H(t,1)  = size(find(z>=k & z<k+binWidth),1);    % find the
                                         % number of points in each bin
    binLocation(t,1)  = k+binWidth/2;                  % assign the
                                         %  center of each bin to an array
    t = t+1;
end
```

Example 22.7. Generate the following two profiles: (a) a sinusoidal profile z_1 of amplitude 2 μm and wavelength 0.8 mm superimposed with uniform noise generated in the interval 0–0.2 μm, and (b) z_2 as the absolute values of z_1, shifted to zero mean. Plot the profiles and the histograms (use 50 bins). Interpret the histograms.

Profile z_1 and z_2 and their histograms are generated as shown below. For clarity, the histogram is plotted so that the vertical axis of the histogram plot is the bin location and it coincides with the profile's vertical axis. The horizontal axis of the histogram plot is the number of profile points in any bin. From Fig. 22.4, it can be seen that profile z_1 is symmetric about the mean and therefore the histogram is also symmetric about the x axis. There are more profile points near the extremities than near the mean, and the histograms reflect this. Profile z_2 is skewed towards the negative side, yet there are more profile points near the upper extremity. The histogram shows this clearly with the peak occurring between 0 and 0.5 μm, while it shows a long tail below the x axis. From this histogram, we can conclude that this profile is negatively skewed.

```
% ADF
dx = 0.001;                              % spacing in mm
l = 8;                                   % profile length in mm
n = l/dx;                                % number of profile points
x = (0:dx:l -dx)';
z1 = 1*sin(2*pi.*x/2)+0.2*rand(n,1);% generate profile z1
plot(x,z1);
xlabel('Distance (mm)');
ylabel('Amplitude (\mum)');
```

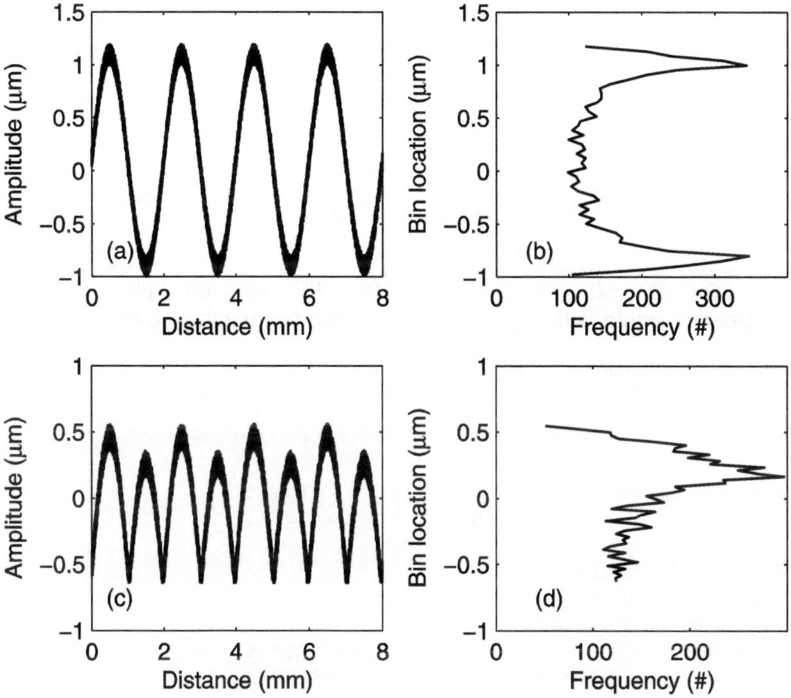

Fig. 22.4 **a** Profile z_1 and **b** its histogram. **c** Profile z_2 and **d** its histogram

```
[H1,binLocation1] = ADF(z1, 50);     % determine ADF using 50 bins
plot(H1,binLocation1);
xlabel('Frequency (#)');
ylabel('Bin location (\mum)');
z2=abs(z1);                          % generate profile z2
z2=z2-mean(z2);                      % shift it to zero mean
plot(x,z2);
xlabel('Distance (mm)');
ylabel('Amplitude (\mum)');
[H2,binLocation2] = ADF(z2, 50);     % determine ADF using 50 bins
plot(H2,binLocation2);
xlabel('Frequency (#)');
ylabel('Bin location (\mum)');
```

Example 22.8. Compute the *BAC* for the profiles in Example 22.7.

If the *ADF* value in any bin is replaced as the sum of the values in that bin and all bins above it, the resulting array is the *BAC*. *BAC* is generally reported as a percentage of profile evaluation length. Continuing from the previous example, the *BAC* can be determined as shown below. Figure 22.5 shows the *BAC* for profiles z_1 and z_2. Notice that the slopes of the *BAC* curves are different; profile z_2 has more material near the upper extremity and therefore the *BAC* has a gentler slope.

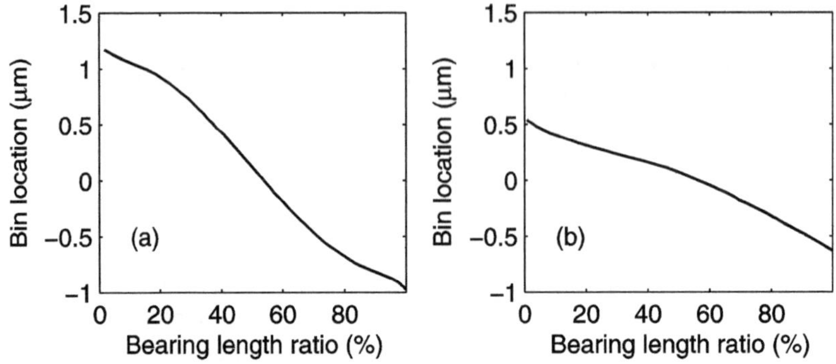

Fig. 22.5a,b Bearing area curves for profiles **a** z_1 and **b** z_2

```
% BAC
for k=size(H1,1):-1:1                          % BAC for z1
    CH1(k,1) = sum(H1(k:size(H1,1),1))*dx*100/l;
end
plot(CH1,binLocation1);
xlabel('Bearing length ratio (%)');
ylabel('Bin location (\mum)');
for k=size(H2,1):-1:1                          % BAC for z2
    CH2(k,1) = sum(H2(k:size(H2,1),1))*dx*100/l;
end
plot(CH2,binLocation2);
xlabel('Bearing length ratio (%)');
ylabel('Bin location (\mum)');
```

Example 22.9. From the *BAC* for profile z_1, determine the height where the bearing length ratio tp is 20%.

Given the two arrays *CH1* and *binLocation1* from Example 22.7, we interpolate between two nearest values around $tp = 20\%$ to determine the height as 0.93 µm.

```
% DETERMINING tp = 20%
u = min(find(CH1 <=20)); % find the index of array CH1 so that
                         % CH1 is just less than 20 %
v=u-1;                   % index of CH1 where CH1 is just greater
                         % than 20 %
Height = (20-CH1(u,1))/(CH1(v,1)-CH1(u,1))*(binLocation1(v,1)-
binLocation1(u,1))+binLocation1(u,1);
```

22.4 Summary

There are several functions that can be computed from surface profiles. These include the *ACV*, the *ACF*, the *PSD*, the *ADF*, the *BAC*, *etc*. These functions are drawn from random process analysis in statistics and serve useful roles in providing additional parameters that can be used to monitor the manufacturing process or relate to part functionality.

Exercises

22.1. Modify the *ACV* implementation in Example 22.1 to compute the "unbiased right-side" *ACV* function given in Eq. 22.2. Plot the resulting *ACV* for profile z given in Example 22.2.

22.2. Modify the *ACV* implementation in Example 22.1 to compute the "unbiased right- and left-side" *ACV* function given as a combination of Eqs. 22.2 and 22.3. Plot the resulting *ACV* for profile z given in Example 22.2 and explain the plot.

22.3. Perform a weighted sum of profile z in Example 22.2 with a copy of itself. How does this relate to any *ACV* function that can be determined from the profile? Explain.

22.4. For a continuous function $z = \sin(2\pi f_o x)$, $0 \leqslant x < l$, evaluate the *PSD* given by $PSD(f) = \frac{1}{l}\left|zf(f)\right|^2$, where zf is the Fourier transform of z, given by $zf(f) = \int_{-\infty}^{\infty} z(x)\,e^{-i2\pi xf}\,dx$. f_o and f are spatial frequencies (hint: the integral is non-zero only when $f = f_o$, and at that location should have a value of $l/2$).

22.5. Which of the profiles z_2 or z_3 in Example 21.4 (Chap. 21) would be more suitable for load carrying purposes? Justify.

References

American Society of Mechanical Engineers 2002, *ASME B46.1-2002, Surface Texture (Surface Roughness, Waviness and Lay)*, ASME. New York, NY, USA.

Peklenik, J. 1967, 'Investigation of the surface typology', *Annals of the CIRP*, vol. 15, pp. 381–385.

Whitehouse, D.J. 1994, *Handbook of Surface Metrology*, Institute of Physics Publishing. Bristol, UK, Philadelphia, USA.

Chapter 23
3D Surface Texture Parameters

23.1 Introduction

There has been a renewed interest in 3D surface texture measurement and analysis in recent years. Several parameters have been identified and studied by researchers, and their efficacy in functional applications has been demonstrated. A number of these parameters are simply 3D extensions of the corresponding 2D parameters. Some, however, are unique to 3D, such as surface texture direction. In the 1990s, an original set of 14 3D parameters were proposed; these were referred to as the "Birmingham 14" (Blunt and Jiang 2003). Subsequently, others were added to the list and they are now in the process of being incorporated into an upcoming international Standard. The ASME B46.1 (American Society of Mechanical Engineers 2002) has included some 3D parameters in its Standard. Several 3D parameters are currently being considered in draft international Standards. In this chapter, we briefly review some of these parameters and illustrate their computations using examples.

23.2 Amplitude and Shape Parameters

The amplitude and shape parameters described in Chap. 21 can be extended for 3D surfaces quite easily. The ASME B46.1 (American Society of Mechanical Engineers 2002) Standard defines amplitude parameters such as average roughness (Sa), maximum area peak height (Sp), minimum area valley depth (Sv), and root mean square roughness (Sq).

The average roughness parameter is defined as

$$Sa = \frac{1}{nx.ny} \sum_{i=1}^{nx} \sum_{j=1}^{ny} |z(j,i)| \,, \tag{23.1}$$

where $z(j,i)$ is the ordinate of the point at the jth row and ith column, and nx and

ny are the number of points along the two directions. Sq is defined as

$$Sq = \sqrt{\frac{1}{nx.ny} \sum_{i=1}^{nx} \sum_{j=1}^{ny} z(j,i)^2} \, . \tag{23.2}$$

Shape parameters such as skew and kurtosis are also an extension of their 2D definitions to 3D. They are given by

$$Sku = \frac{1}{Sq^4} \frac{1}{nx.ny} \sum_{i=1}^{nx} \sum_{j=1}^{ny} z(j,i)^4 \tag{23.3}$$

$$Ssk = \frac{1}{Sq^3} \frac{1}{nx.ny} \sum_{i=1}^{nx} \sum_{j=1}^{ny} z(j,i)^3 \, . \tag{23.4}$$

As mentioned before, our focus here is on computational aspects of some of these parameters, and we have therefore not cataloged every parameter conceivable.

Example 23.1. Write a MATLAB function that accepts a surface z and returns the amplitude parameters Sa, Sq, Sp, Sv, and St in that order, computed on z and evaluated over the evaluation area.

```
function [Sa, Sq, Sp, Sv, St] = myAmplitudeParameters(z)
% number of rows in z corresponds to number of points along the y
% direction and number of columns to points along x.
ny = size(z,1);                      % determine length of z along y
nx = size(z,2);                      % determine length of z along x
z = z - mean(mean(z,1),2);           % shift profile to zero mean
Sa = sum(sum(abs(z)))/nx/ny;         % compute Sa
Sq = sqrt(sum(sum(z.*z))/nx/ny);     % compute Sq
Sp = max(max(z));                    % compute Sp
Sv = abs(min(min(z)));               % compute Sv
St = Sp + abs(Sv);                   % compute St
```

23.3 *AACV*, *APSD*, and *BAC* for 3D Surfaces

The area autocovariance function (*AACV*) for 3D surfaces is defined as

$$AACV(j',i') = \frac{1}{nx.ny} \sum_{i=1}^{nx-i'} \sum_{j=1}^{ny-j'} z(j,i)z(j+j',i+i') \, , 0 \leqslant j \leqslant ny-1 \, , 0 \leqslant i \leqslant nx-1 \, . \tag{23.5}$$

The area power spectral density function (*ASPD*) is given by

$$APSD(fx,fy) = \frac{\Delta x.\Delta y}{nx.ny} \left| zf(fx,fy) \right|^2 \, , \tag{23.6}$$

where zf is the 2D DFT of the surface z.

The bearing area ratio is defined as the ratio of the area of intersection of the measured topography with a selected surface parallel to the mean surface to the evaluation area, and is expressed as a percentage. We illustrate these in the following examples.

Example 23.2. Write a MATLAB function to generate the *AACV* function as given by Eq. 23.5.

```
function [shiftx, shifty, AACV] = computeAACV(z,dx,dy)
nx = size(z,2);
ny = size(z,1);
for ty = 1:ny
    for tx = 1:nx
        temp = 0;
            for j =1:ny-ty
                for i=1:nx-tx
                temp = temp + z(j,i)*z(j+ty-1,i+tx-1);
            end
        end
        AACV(ty,tx) = temp;
    end
end
AACV = AACV/nx/ny;
shiftx = (1:nx)*dx;
shifty = (1:ny)*dy;
```

Example 23.3. For the 3D surface in Example 8.1, compute and plot the *AACV* function.

The *AACV* function is plotted in Fig. 23.1. Because the surface Z is periodic in only one direction, the *AACV* is also periodic in that direction only. Along the other direction, the *AACV* falls to zero because the overlap area decreases.

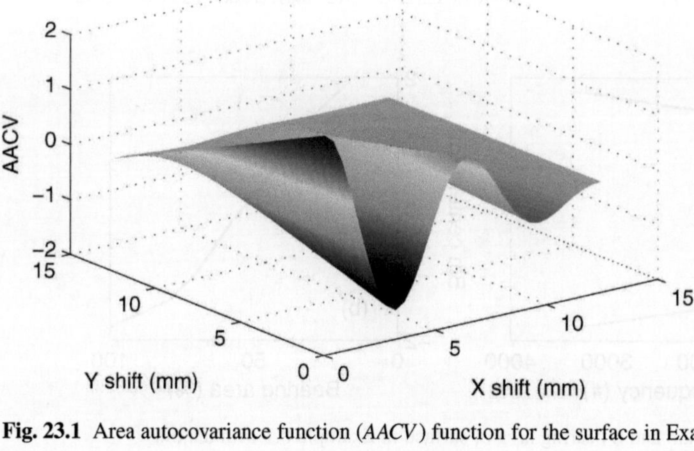

Fig. 23.1 Area autocovariance function (*AACV*) function for the surface in Example 8.1

```
[shiftx,shifty,AACV] = computeAACV(z,0.1,0.1);
surf(shiftx,shifty,AACV)
shading interp;
xlabel('X shift (mm)');
ylabel('Y shift (mm)');
zlabel('AACV');
```

Example 23.4. Compute the *APSD* function for the same surface.

```
dx = 100;                          % spacing in um
dy = 100;
zf = fft2(z);                      % evaluate DFT of z
PSD = dx*dy/nx/ny*(abs(zf).^2);
jx = (1:1:floor(nx/2)+1)';
freqx = (jx-1)/(nx*dx);            % generate frequency array
jy = (1:1:floor(ny/2)+1)';
freqy = (jy-1)/(ny*dy);            % generate frequency array
surf(freqx,freqy,PSD(1:65,1:65))   % plot PSD as
                                   % function of frequency
```

The peak occurs at position $(1,3)$, that is, at a frequency of $0.1563 \times 10^{-3}\,\mu\text{m}^{-1}$ or a wavelength of $1/(0.1563 \times 10^{-3}) = 6.4$ mm (`freqx`(3)) along x and zero along y.

Example 23.5. Determine the histogram and bearing area of the 3D surface z in Example 8.1.

The histogram H and bearing area CH can be computed and plotted as shown below (see Fig. 23.2).

```
[CH,H,binLocation] = AADF(z, 10);
plot(H,binLocation)
plot(CH,binLocation)
```

where the `AADF` function is given as

```
function [CH,H,binLocation] = AADF(z, numBins)
a = max(max(z));              % determine the maximum value in z
```

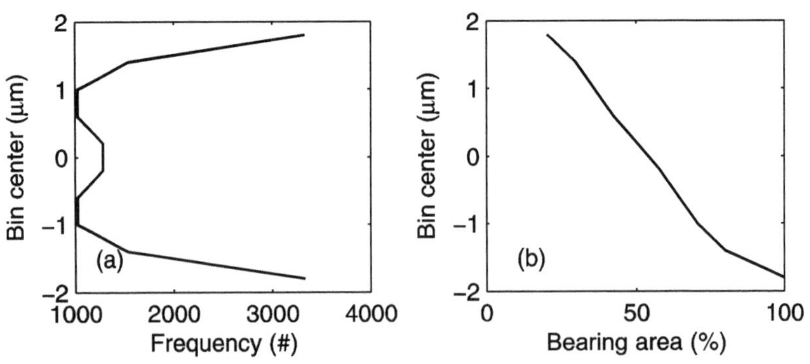

Fig. 23.2 **a** Histogram and **b** bearing area of surface in Example 8.1 with ten bins

```
b = min(min(z));            % determine the minimum value in z
binWidth = (a-b)/numBins; % determine bin width
t = 1;
for k=b:binWidth:a-binWidth
    H(t,1) = size(find(z>=k & z<k+binWidth),1); % find the number
                            % of points in each bin
    binLocation(t,1) = k+binWidth/2; % assign the center
                            % of each bin to an array
    t = t+1;
end
for k=size(H,1):-1:1        % BAC for z
    CH(k,1) = sum(H(k:size(H,1),1))*100/size(z,1)/size(z,2);
end
```

23.4 Spacing and Hybrid Parameters

There are a number of spacing and hybrid parameters that are defined in the upcoming ISO 25178-2 international Standard due to be released in 2009. *Sal* is the horizontal distance of the autocorrelation function that has the fastest decay in any direction to a specified threshold. Continuing from Example 23.3, we scan the *AACV* radially from the origin in all directions to find the shortest distance to the point where the *AACV* function drops to, say, 0.1. This is shown in Example 23.6.

Example 23.6. Compute *Sal* for the *AACV* data in Example 23.3.

```
Threshold = 0.1;
U = find(AACV <= Threshold);
Distance = sqrt(U.^2);
Sal = min(Distance)
```

Summit density (*Sds*) is the number of summits per unit area making up the surface. Summits are derived from peaks. A peak is defined as any point that rises above eight nearest neighbors. Peaks have to be separated by a certain distance threshold (at least 1% of the measured distance along x or y). Additionally, summits are only found above a height threshold (say, 5% of Sz above the mean plane). An algorithm to compute summit density is as follows:

1. Identify all points (say, m points in all) that lie above the height threshold and order them in some sequence from 1 to m.
2. Define an array called *Peak* that contains locations of peak points.
3. For point $i = 1$ to m:

    a. If point i is larger than its eight nearest neighbors and if the distance of point i from all points in *Peak* array is greater than the distance threshold, then assign the point as a peak.
    b. Else increment i and proceed to step 2.

We leave it as an exercise to the reader to implement the above algorithm as a program in MATLAB.

23.5 Summary

In this chapter, we have described some 3D surface texture parameters and illustrated their computational aspects. Some surface parameters are simply the 3D counterparts to the 2D parameters, and are reasonably simple to compute. A new international Standard, currently in draft stage, incorporates several additional new parameters that exploit the true 3D nature of surfaces.

Exercises

23.1. Write a MATLAB function to determine *Sku* and *Ssk* for any 3D surface z.

23.2. Generate a 3D surface with 128×128 points that is only composed of random noise (mean $= 0$ and standard deviation $= 1$ μm). Determine *Sku* and *Ssk* for this surface.

23.3. For the surface in Exercise 8.3 (12.8 mm long, sampling interval of 0.1 mm, and a sinusoid of amplitude 1 μm and wavelength 6.4 mm in both directions x and y), without computing the *APDF*, determine the location of the peaks in the *APDF* array. Then compute the *APDF*, plot it, and verify that the peaks are at the correct location.

23.4. Compute and plot the *AACV* of the surface in Exercise 8.3. Explain the shape of the *AACV* plot.

23.5. For the surface in Exercise 23.2, compute and plot the BAC and the histogram. What does the histogram indicate?

References

American Society of Mechanical Engineers 2002, *ASME B46.1-2002, Surface Texture (Surface Roughness, Waviness and Lay)*, ASME. New York, NY, USA.

Blunt, L. and Jiang, X. (eds) 2003, *Advanced Techniques for Assessment Surface Topography: Development of a Basis for 3D Surface Texture Standards*, Kogan Page Science. London, UK.

Part V
Errors and Uncertainty

In Part V, we review how uncertainty in measurements impact computations performed in surface and roundness metrology. In Chap. 24, we provide a brief introduction leading to the topic of uncertainty propagation in computations. We elaborate on this further in Chap. 25, and illustrate how this affects the three topics we have discussed earlier: fitting, filtering and parameterization. In Chap. 26, we review some interesting mathematical techniques used in error separation in roundness metrology. In Chap. 27, we present a short overview of some emerging topics in computational surface and roundness metrology that we have not covered in this text.

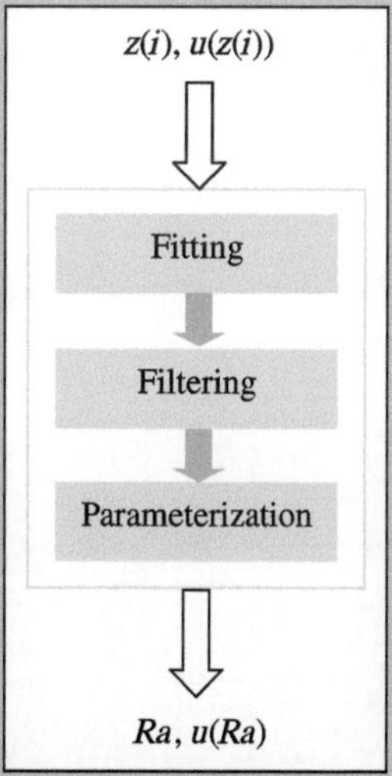

Fig. V Profile points $(z(i))$ along with their uncertainty $(u(z(i)))$ as input to the chain of computations. The uncertainty propagates through the computations. The outputs are the parameters of interest (such as, Ra) and the uncertainty in the parameters $(u(Ra))$

Chapter 24
Uncertainty Considerations

24.1 Introduction

In the preceding chapters, we focused on techniques such as fitting, filtering and parameterization that aid in the determination of the value of a quantity (the measurand) we are interested in, say *Ra*. These techniques involved algebraic manipulations of the raw data without any considerations of the uncertainty in that data. In reality, there is always some uncertainty associated with the raw data and this uncertainty impacts the measurand. We study this topic here. The ISO Guide to the Expression of Uncertainty in Measurement (GUM) (International Organization for Standardization 1995) and the corresponding US Guide (American National Standards Institute 1997) are definitive references.

24.2 Random and Systematic Components

Measurement uncertainty can arise from a number of sources such as the probing system, environment, operator influences, finite sampling *etc*. Some components of uncertainty in a measurement are random in nature. That is, they vary from measurement to measurement and therefore cannot not be predicted or compensated. For example, electrical noise in the probing system of a stylus instrument may be random in nature. Ten consecutive readings of the probe at the same location will result in ten different readings.

There are other sources that impact every measurement in an identical manner. These components, referred to as systematic components, can be compensated for. Systematic components introduce a bias into the measurement. A careful analysis of the uncertainty sources usually leads to the identification, and therefore compensation, of such sources. Because the compensation for systematic errors is never perfect, there is always a residual uncertainty after compensation.

24.3 Uncertainty Modeling

According to the GUM, uncertainties sources are modeled using probability distributions. The uncertainty due to any source is assigned a value equal to the standard deviation of the chosen probability distribution. Experiments and experience dictate the type of distribution and its parameters. The normal distribution is the most commonly assigned distribution for most sources. For example, 100 repeated measurements at the same point on a surface may yield values clustered around a mean of say, 1.3 μm. The spread (standard deviation) on either side of the mean may be determined to be 0.1 μm. The quantity may therefore be modeled using a normal distribution of 1.3 μm mean and standard deviation 0.1 μm. The mean is the nominal value of the quantity and the standard deviation is its uncertainty. Such experimental determination of the standard deviation is referred to as Type A uncertainty determination. In some instances, experience may dictate the chosen distribution and its parameters. The GUM refers to this as Type B determination of standard uncertainty.

The first step in developing an uncertainty budget is the identification of every source of uncertainty and the assignment of appropriate distributions. We are not concerned with this aspect of uncertainty modeling in this text. After all uncertainty components have been assigned a standard uncertainty value, they have to be combined in some way so that we may assess their impact on the measurand. We focus on this aspect here.

24.4 Uncertainty Propagation

Any measurement uncertainty in the raw data propagates through the chain of computations that are performed to impact final parameters of interest. In addition, the algorithms and computations that massage the raw data may introduce new errors as well; these also impact the final parameters of interest (Hopp 1993). Therefore, an understanding of uncertainty propagation in computations is critical to estimating uncertainty on final parameters. In the context of surface texture and roundness analysis, the problems we are concerned with involve studying how the uncertainty in each profile point propagates through fitting and filtering operations to impact the measurand (such as *Ra*).

The GUM provides a technique for propagating uncertainties. We briefly review that in Chap. 25. We also review a technique referred to as Monte Carlo simulation (MCS) that achieves the same objective as the GUM, but using numerical simulation.

24.5 Systematic Errors: an Example

According to the GUM, systematic components in a measurement are not treated any differently from random components. All systematic errors are identified and corrected for. Any residual terms are modeled using probability distributions and the standard uncertainties are propagated through the chain of computations.

Roundness metrology offers an interesting example for this case and we discuss that in Chap. 26. Precision spindles used in roundness measurements have undesired radial motion. This radial motion convolves with the part surface and therefore yields a roundness profile that has both part and spindle radial errors in it. Some of these errors are due to a random component in the precision spindle's radial motion and cannot be removed. Some portion of the precision spindle's radial motion is however systematic; that is, this component repeats from measurement to measurement and can therefore be identified and removed. There are several interesting mathematical techniques that are involved here and we discuss these in that chapter. We also discuss how uncertainty in each raw profile point impacts the error separation process and therefore the final profile points.

References

American National Standards Institute 1997, *ANSI/NCSL Z540-2-1997, U.S. Guide to the Expression of Uncertainty in Measurement*, ANSI. Boulder, CO, USA.

Hopp, T. 1993, 'Computational metrology', *Manufacturing Review*, vol. 6, no. 4., pp. 295–304.

International Organization for Standardization 1995, *ISO/IEC Guide 98:1995, ISO Guide to the Expression of Uncertainty in Measurement*, ISO. Geneva, Switzerland.

Chapter 25
Uncertainty Propagation in Computations

25.1 Introduction

In any measurement, the desired output quantity is rarely directly measured, but instead is a function of several input quantities that are measured. For example, *Ra* (the desired output quantity) is not directly measured, but is a function of several profile points (the inputs) that are measured using surface finish instrumentation. The input quantities have some uncertainty associated with them and this uncertainty propagates through the chain of computations to impact the output quantity.

Let the functional relationship between Y (the output quantity) and $X(i)$ (the ith input quantity) be given by

$$Y = f(X(1), X(2), \ldots, X(n)) , \tag{25.1}$$

where n is the number of inputs. The distribution of possible values for each input quantity $X(i)$ can either be determined experimentally or theoretically. The standard deviation of this distribution is regarded as the standard uncertainty, $u(X(i))$, of the input quantity $X(i)$.

Then, according to the law of propagation of uncertainty, the standard uncertainty in Y, $u(Y)$, is given by

$$u(Y) = \sqrt{\frac{\partial^2 f}{\partial X(1)^2} u^2(X(1)) + \frac{\partial^2 f}{\partial X(2)^2} u^2(X(2)) + \ldots + \frac{\partial^2 f}{\partial X(n)^2} u^2(X(n))} \, . \tag{25.2}$$

Example 25.1. Let $Y_1 = X(1) + X(2)$, $Y_2 = X(1) - X(2)$, $Y_3 = X(1)X(2)$, and $Y_4 = X(1)/X(2)$. Let nominal values for $X(1)$ and $X(2)$ be drawn from a normal distribution of mean 2 and 0.1 standard deviation. What are the uncertainties in Y_1, Y_2, Y_3, and Y_4?

Using equation Eq. 25.2:

$$u(Y_1) = \sqrt{u^2(X(1)) + u^2(X(2))} = \sqrt{0.1^2 + 0.1^2} = 0.14$$
$$u(Y_2) = \sqrt{u^2(X(1)) + u^2(X(2))} = 0.14$$
$$u(Y_3) = \sqrt{X(2)^2.u^2(X(1)) + X(1)^2.u^2(X(2))} = 0.28$$
$$u(Y_4) = \sqrt{(1/X(2)^2).u^2(X(1)) + (X(1)^2/X(2)^4).u^2(X(2))} = 0.07$$

25.2 Intervals at Some Level of Confidence

In addition to obtaining an estimate of the standard uncertainty of the output quantity, we often have to specify an interval for the output quantity at some level of confidence. The interval defines a bound on Y at some level of probability, say p, so that a large fraction p of the estimates for Y will be within that bound. The larger the required level of confidence, the larger will be the interval defined by the bounds. The standard deviation alone is not sufficient to estimate confidence intervals for Y; rather, we also require knowledge of the distribution itself. If Y were normally distributed, the interval corresponding to 95% probability (level of confidence) is approximately $\pm 2u(Y)$, and the interval corresponding to 99% is $\pm 3u(Y)$ [see definition for normal distribution (Kreyzig 1999)].

Example 25.2. Let an output quantity Y_1 be normally distributed with a standard uncertainty $u(Y_1) = 0.1$ determined from the law of uncertainty propagation. Let another output quantity Y_2 belong to a rectangular distribution with the same standard uncertainty $u(Y_2) = 0.1$. Let the mean values for Y_1 and Y_2 be zero. Determine the interval at 95% confidence in each case.

The interval for the normal distribution at 95% confidence is determined as approximately twice the standard deviation, hence -0.2 to 0.2. A rectangular distribution of half-width a has a standard deviation of $a/\sqrt{3}$ (Kreyzig 1999). Therefore, if $a/\sqrt{3} = 0.1$, $a = 0.17$. The 95% confidence is then given by -0.17×0.95 to 0.17×0.95, or -0.16 to 0.16.

Example 25.3. What are the intervals at 95% confidence for Y_1, Y_2, Y_3, and Y_4 in Example 25.1?

Assuming the output quantities are normally distributed, the intervals at 95% confidence are

$$Y_1: 4 \pm 0.28$$
$$Y_2: 0 \pm 0.28$$
$$Y_3: 4 \pm 0.56$$
$$Y_4: 1 \pm 0.14$$

25.3 The Central Limit Theorem

The combined standard uncertainty as determined by the law of uncertainty propagation is sufficient in most applications. However, in some situations there is an additional requirement to know the interval at some level of confidence and, as mentioned before, we require knowledge of the distribution of the output quantity for this purpose. In many situations, we simply approximate the distribution of the output quantity Y as the normal distribution and then proceed to establish the interval of confidence. This assumption of normality is not arbitrary; rather, it is the consequence of the central limit theorem which we discuss here.

We stated earlier that

$$Y = f(X(1), X(2), \ldots, X(n)) \,. \tag{25.3}$$

If Y can be approximated by a first-order Taylor series expansion so that it is linear in the input quantities as shown:

$$Y = c(1)X(1) + c(2)X(2) + \ldots + c(n)X(n) \,, \tag{25.4}$$

then, if each $X(i)$ belonged to a normal distribution, the output quantity Y can also be shown to be normally distributed. Note that $c(i)$ are coefficients from the Taylor series expansion. The variance of Y can then be given by

$$u^2(Y) = c(1)^2u^2(X(1)) + c(2)^2u^2(X(2)) + \ldots + c(n)^2u^2(X(n)) \,. \tag{25.5}$$

The central limit theorem provides a valuable extension to the preceding formulation. It says that Y can be approximated by a normal distribution even if some or all of the $X(i)$ are not normally distributed, provided the variance of $Y$is larger than each of the variances of the components $(c(i)^2u^2(X(i)))$ of non-normally distributed $X(i)$. A corollary of the central limit theorem is that the average of n samples drawn from a non-normal distribution tends to be normally distributed when n is large.

The central limit theorem is useful in a large number of situations where the function f can be approximated by a first-order Taylor series expansion, and sufficient averaging of the input quantities (especially of non-normal distributions) produces output quantity Y that is approximately normal. Therefore, in real-life situations, Y can be approximated by a normal distribution and the interval of confidence therefore determined.

Example 25.4. Verify the central limit theorem in case of a random variable belonging to a rectangular distribution with bounds -1 to 1 by showing that the average of n samples drawn from this distribution tends to become a normal distribution as n tends to infinity.

Let random variable A be drawn from a rectangular distribution. Random variable B is obtained by averaging two samples from a rectangular distribution. C is obtained by averaging 50 samples, and D by averaging 100 samples. Note that the histogram of *A* is indeed rectangular as expected. *B* is triangular, *C* and *D* are nearly normal (Fig. 25.1).

```
for i=1:5000
    A(i) = (rand(1,1)-0.5)/0.5;
    B(i) = (mean(rand(1,2))-0.5)/0.5;
    C(i) = (mean(rand(1,50))-0.5)/0.5;
    D(i) = (mean(rand(1,100))-0.5)/0.5;
end
[xa,ca] = hist(A,20);[xb,cb] = hist(B,20);[xc,cc] = hist(C,20);
[xd,dd] = hist(D,20);
bar(ca,xa)
bar(cb,xb)
bar(cc,xc)
bar(dd,xd)
```

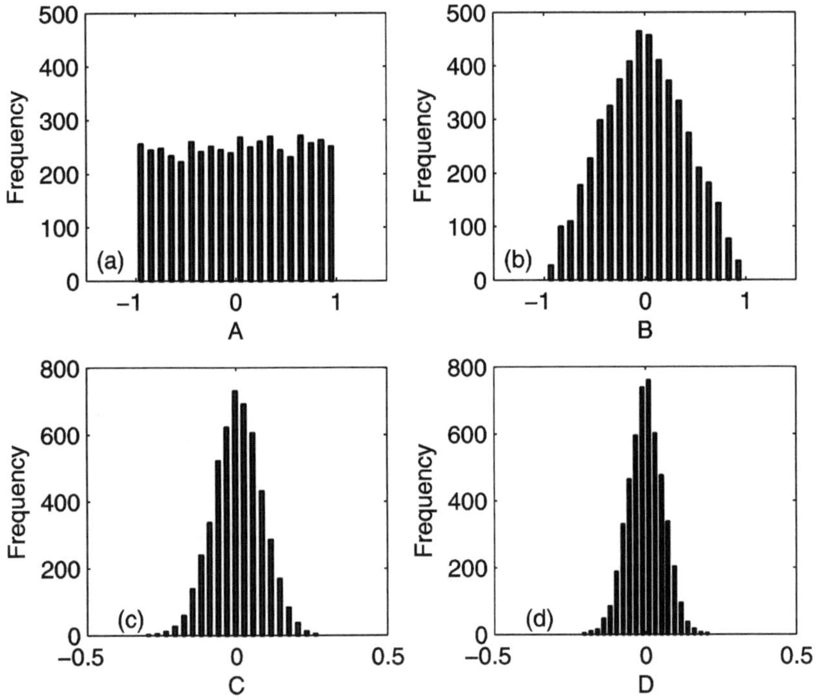

Fig. 25.1a–d Verifying the central limit theorem. **a** Random variable *A* belonging to a rectangular distribution. **b** Random variable *B* defined as the average of two samples from the rectangular distribution appears to belong to a triangular distribution. **c** Random variable *C* defined as the average of 50 samples. **d** Random variable *D* defined as the average of 100 samples from the rectangular distribution. *D* appears to be normally distributed

25.4 Limitations of GUM

The primary limitation of GUM in the context of uncertainty propagation is that we require a functional relationship between the output and input quantities that can be approximated by a first-order Taylor series expansion. In many of the data processing techniques we have seen in previous sections, we can only express this relationship in the form of a computer program, not as an analytic function. The Monte Carlo method is a numerical simulation technique that overcomes this problem. We discuss this next.

25.5 The Monte Carlo Method

The idea behind the Monte Carlo simulation (MCS) (Cox *et al.*, 2001) is to numerically propagate the entire distribution instead of just the parameter of a distribution (standard deviation) as in the GUM. The method works as follows.

Each input quantity is described by a distribution as mentioned before. Values are sampled from the distribution for each input quantity, and the output quantity is calculated. This process is repeated several times to yield a histogram of the set of values for the output quantity. This histogram not only provides information on the standard deviation but also yields the distribution itself. The standard deviation is a measure of the uncertainty in Y. The coverage interval can also be determined from the set of values for Y as follows: the values for Y are sorted in ascending order and the interval that accommodates the central p fraction of values is the interval of confidence corresponding to the probability p.

Example 25.5. Determine the standard uncertainty and the 95% confidence interval using MCS for Y_1, Y_2, Y_3, and Y_4 in Example 25.1.

```
X(1) = 2;
X(2) = 2;
for i=1:50000
    a = normrnd(0,0.1);
    b = normrnd(0,0.1);
    Y1(i) = (X(1)+a) + (X(2)+b);
    Y2(i) = (X(1)+a) - (X(2)+b);
    Y3(i) = (X(1)+a)*( X(2)+b);
    Y4(i) = (X(1)+a)/( X(2)+b);
end
[std(Y1) std(Y2) std(Y3) std(Y4)]
```

The standard deviations are 0.14, 0.14, 0.28, and 0.07 and are identical to that obtained earlier.

```
Y1=sort(Y1); Y2=sort(Y2); Y3=sort(Y3);Y4=sort(Y4);
a = 50000*2.5/100;b=50000*97.5/100;
[Y1(a) Y1(b)]
[Y2(a) Y2(b)]
```

```
[Y3(a) Y3(b)]
[Y4(a) Y4(b)]
```

$$Y_1: \quad 3.73 \text{ to } 4.28$$
$$Y_2: -0.28 \text{ to } 0.28$$
$$Y_3: \quad 3.46 \text{ to } 4.57$$
$$Y_4: \quad 0.87 \text{ to } 1.15$$

Note that the intervals of confidence are identical to that obtained in Example 25.3.

Example 25.6. In Example 25.1, assume $X(1)$ and $X(2)$ can have any value in the range 2 ± 0.1732 with equal probability. Are the distributions for Y_1, Y_2, Y_3, and Y_4 normal? If not, suggest how the central limit theorem may be exploited to obtain normal distributions for Y_1, Y_2, Y_3, and Y_4.

According to the GUM, the standard uncertainties of $X(1)$ and $X(2)$ are $0.1732/\sqrt{3} = 0.1$. Propagating this, we get the results shown in Example 25.1. MCS can be performed as shown below.

```
for i=1:50000
    a = (rand(1)-0.5)/0.5*0.1732;
    b = (rand(1)-0.5)/0.5*0.1732;
    Y1(i) = (X(1)+a) + (X(2)+b);
    Y2(i) = (X(1)+a) - (X(2)+b);
    Y3(i) = (X(1)+a)*(X(2)+b);
    Y4(i) = (X(1)+a)/(X(2)+b);
end
[x,c] = hist(Y1,100); bar(c,x)
[x,c] = hist(Y2,100); bar(c,x)
[x,c] = hist(Y3,100); bar(c,x)
[x,c] = hist(Y4,100); bar(c,x)
```

The distributions are triangular. Averaging the inputs as in Example 25.4 will result in Y being closer to a normal distribution.

25.6 Filtering Surface Profiles

In this section, we consider some aspects involved in uncertainty propagation in filtering surface profiles. Consider a profile z containing ten points. Let the uncertainty in the ith point be $u(z(i))$. Further because the measurement process is uniform, assume $u(z(i))$ is identical for all points and is denoted by $u(z)$. Let each point in z be assigned a normal distribution of 0.1 μm standard deviation. Let the profile z be filtered by a rectangular filter S containing three elements (each element is therefore equal to $1/3$). Assume that there is no uncertainty involved in the determination of the filter itself and also that there is no uncertainty in the determination of x position of the point. Using convolution, we know that the waviness profile is determined as the weighted average of the profile with the filter. The waviness profile at the ith location $w(i)$ is given by

$$w(i) = S(1)z(i-1) + S(2)z(i) + S(3)z(i+1)\,. \tag{25.6}$$

The uncertainty in $w(i)$, denoted as $u(w(i))$, is therefore given by the law of propagation of uncertainty as

$$\begin{aligned} u(w(i)) &= \sqrt{S(1)^2u^2(z(i-1)) + S(2)^2u^2(z(i))^2 + S(3)^2u^2(z(i+1))} \\ &= u(z)\sqrt{\sum_{j=1}^{3} S(j)^2} = 0.058\,\mu\text{m}\,. \end{aligned} \tag{25.7}$$

The uncertainty in each point of the waviness profile is smaller in comparison to that of the unfiltered profile. This is only to be expected because the filtering operation is effectively an averaging operation.

If the filter used were a Gaussian filter containing m points, the uncertainty in any point of the waviness profile is

$$u(w(i)) = u(z)\sqrt{\sum_{j=1}^{m} S(j)^2}\,. \tag{25.8}$$

We next look at the uncertainty in the roughness profile. The ith point of the roughness profile $r(i)$ is given by $z(i) - w(i)$. Because we know the standard uncertainties of each point in z and w, it is tempting to use the law of propagation of uncertainty and write the uncertainty in $r(i)$ as $\sqrt{u(z(i))^2 + u(w(i))^2} = 0.116\,\mu$m. However, this equation is only valid when z and w are independent. In this case, the uncertainty in the waviness profile is dependent on the uncertainty of the profile points z, or more specifically, $w(i)$ and $z(i)$ are correlated. In the case of the rectangular filter example, we can easily remove this correlation by expressing $w(i)$ in terms of $z(i-1)$, $z(i)$, and $z(i+1)$ as follows:

$$\begin{aligned} r(i) &= z(i) - w(i) \\ &= z(i) - (S(1)z(i-1) + S(2)z(i) + S(3)z(i+1)) \\ &= (2/3)z(i) - (1/3)z(i-1) - (1/3)z(i+1)\,. \end{aligned} \tag{25.9}$$

The uncertainty of the roughness profile at the ith location is then given by

$$\begin{aligned} u(r(i)) &= \sqrt{(4/9)u^2(z(i)) + (1/9)u^2(z(i-1)) + (1/9)u^2(z(i+1))} \\ &= 0.1\sqrt{(4/9) + 1/9 + 1/9} \\ &= 0.082\,\mu\text{m}\,. \end{aligned} \tag{25.10}$$

Notice that this uncertainty is smaller than the uncertainty in the input profile z. Again, this is only to be expected because the total uncertainty in z is distributed between r and w.

It may not be possible to eliminate correlation between inputs in every situation. More generally, in the presence of correlation in the input quantities, the law of propagation of uncertainty can be stated as

$$u^2(Y) = \sum_{i=1}^{n} c(i)^2 u^2(X(i)) + 2\sum_{i=1}^{n-1}\sum_{j=i+1}^{n} u(X(i))u(X(j))cr(X(i),X(j)) \,, \quad (25.11)$$

where $c(i)$ is the partial differential $\frac{\partial f}{\partial X(i)}$ and $cr(X(i),X(j))$is the correlation coefficient between $X(i)$ and $X(j)$.

Krystek (2001) provides an analytic expression for the uncertainty in roughness and waviness profiles when using a Gaussian filter. But it is evident that the computations shown above can become quite cumbersome for even some simple situations not discussed here. MCS is then a viable tool for estimating the effects of uncertainty propagation. In the following example, we illustrate the concepts discussed here.

Example 25.7. Let z be a profile that is filtered using a rectangular filter with three elements (each of value $1/3$). The uncertainty in the roughness and waviness profiles were computed in Sect. 25.6. Verify those values using MCS.

Consider three points on the profile z that the filter window currently envelops. The central point $z(2)$ is the point of interest because the waviness (w) and roughness (r) profile values are computed for this location. Note that the nominal values of the profile points are not of interest (hence assumed to be zero); only the uncertainties are.

```
S = [1 1 1]'/3;
z = [0 0 0]'; % nominal profile points are zero.
   for i=1:5000
% Note that each column in Z represents a draw from a normal
% distribution. Each row represents a profile point at a given
% location. Thus, Z(:,31) represents the three profile points in
% the 31st draw from the distribution
   Z(:,i) = z + normrnd(0,0.1*ones(3,1));
   C = sum(S.*Z(:,i));
   w(i) = C;
   r(i) = Z(2,i)-C;
end
[std(w) std(r)]
```

The standard uncertainties for w and r are evaluated to be 0.058 μm and 0.081 μm, which are in agreement with those computed earlier.

Example 25.8. Compute the uncertainty in the roughness profile by using Eq. 25.11.

Continuing from the previous example, we determine the correlation coefficient of $z(2)$ and w as shown

```
cr = corrcoef([Z(2,:)' w']); % note that cr is of size 2 x 2.
cr = cr(1,2);
```

The value for cr is 0.57. The functional relationship between the desired output quantity r and the input quantities $z(2)$ and w is given by $r = z(2) - w$. The variance computed using the law of propagation of uncertainty in Eq. 25.11, is

$$\begin{aligned} u^2(r) &= u^2(z(2)) + u^2(w) - 2u(z(2))u(w)cr \\ &= 0.1^2 + 0.058^2 - 2(0.058 \times 0.1)(0.57) = 0.0068 \end{aligned}$$

The standard uncertainty $u(r) = 0.082\,\mu\text{m}$.

25.7 Fitting Substitute Geometry

We focus our attention on uncertainty propagation in fitting best-fit geometry in this section. For the simple case of fitting a line to a set of points, we show how the law of propagation of uncertainty can be used to determine the uncertainty in the estimates for the line. In examples, we show how MCS can be used for the same purpose. For more complex geometries, MCS may be a more viable tool to determine estimates of uncertainty in parameters. Hopp (1993) presents a comprehensive treatment of uncertainty propagation in fitting.

Let the best-fit line be defined as $z = a + bx$, where a is the intercept and b is the slope. Recall from Chap. 15 that

$$\begin{aligned} b &= \frac{\sum_{i=1}^{n} x(i) \sum_{i=1}^{n} z(i) - n \sum_{i=1}^{n} x(i)z(i)}{\left(\sum_{i=1}^{n} x(i)\right)^2 - n \sum_{i=1}^{n} x(i)^2} \\ a &= \frac{\sum_{i=1}^{n} x(i) \sum_{i=1}^{n} x(i)z(i) - \sum_{i=1}^{n} x(i)^2 \sum_{i=1}^{n} z(i)}{\left(\sum_{i=1}^{n} x(i)\right)^2 - n \sum_{i=1}^{n} x(i)^2} \end{aligned} \quad . \tag{25.12}$$

Let the standard uncertainty in each measured point $z(i)$ be denoted by $u(z(i))$. Then, the uncertainty in the slope can be derived as follows (Draper and Smith 1998):

$$b = \frac{\sum_{i=1}^{n} x(i) \sum_{i=1}^{n} z(i) - n \sum_{i=1}^{n} x(i)z(i)}{\left(\sum_{i=1}^{n} x(i)\right)^2 - n \sum_{i=1}^{n} x(i)^2} = c(1)z(1) + c(2)z(2) + \ldots + c(n)z(n)\,, \tag{25.13}$$

where

$$c(i) = \frac{x(i) - \bar{x}}{\sum_{i=1}^{n} (x(i) - \bar{x})^2} \tag{25.14}$$

and $\bar{x}$ is the average of x. Using the law of propagation of uncertainty, the variance in b can be expressed as

$$u^2(b) = c(1)^2u^2(z(1)) + c(2)^2u^2(z(2)) + \ldots + c(n)^2u^2(z(n)) \,. \tag{25.15}$$

If the uncertainties in the measured points are all equal and denoted by $u(z)$, then

$$u(b) = u(z)\sqrt{\sum_{i=1}^{n} c(i)^2} = u(z)\sqrt{\frac{1}{\sum\limits_{i=1}^{n}(x(i)-\bar{x})^2}} \,. \tag{25.16}$$

In a similar manner, the uncertainty in a can be shown to be

$$u(a) = u(z)\sqrt{\frac{\sum\limits_{i=1}^{n} x(i)^2}{n\sum\limits_{i=1}^{n}(x(i)-\bar{x})^2}} \,. \tag{25.17}$$

Example 25.9. Consider a surface profile containing 20 points that was measured on an ideally flat surface inclined at 45° with the horizontal. Assume that the uncertainty in each of the points is normally distributed with a standard deviation of 0.1 μm. What is the uncertainty in the parameters of the best-fit line used to remove the tilt from the profile? Compute using Eqs. 25.16 and 25.17, and verify using MCS.

We first compute the uncertainty using Eqs. 25.16 and 25.17.

```
x = (1:20)';
z = x;
ub = 0.1*sqrt(1/sum((x-mean(x)).^2));
ua = 0.1*sqrt(sum(x.^2)/20/sum((x-mean(x)).^2));
[ua ub]
```

Next, we compute the uncertainty using MCS.

```
for i=1:5000
    E = normrnd(0,0.1*ones(20,1));
% Note that each column in Z represents a draw from a normal
% distribution. Each row represents a profile point at a given
% location. Thus, Z(:,31) represents the 20 profile points in
% the 31st draw from the distribution
    Z(:,i) = z + E;
    [P,S]= polyfit(x,Z(:,i),1);
    B(i) = P(1);
    A(i) = P(2);
    % d is the deviation from the best-fit line
    d(:,i) = Z(:,i) - (B(i).*x+A(i));
end
[std(A) std(B)]
```

Both methods yield the same uncertainty of 0.0039 for the slope and 0.0465 µm for the intercept.

Example 25.10. What is the uncertainty in each point of the leveled profile in Example 25.9?

Continuing from the previous example, the uncertainty in the jth point can be obtained as follows using the command `std(d(j,:))`.

We can also use the law of propagation of uncertainty after recognizing the existence of correlation between slope, intercept, and raw profile points. The correlation matrix between all these input quantities can be obtained as shown:

```
cr = corrcoef([Z' A' B']);
```

The functional relationship between the output d and the inputs can be expressed as

$$d(i) = z(i) - a - bx(i)\,.$$

The variance of $d(i)$ can be written as

$$\begin{aligned} u^2(d(i)) =& u^2(z(i)) + u^2(a) + u^2(b)x(i)^2 + 2u(z(i))u(a)(-1)cr(x(i),a) \\ &+ 2u(z(i))u(b)(-1)x(i)cr(z(i),b) + 2u(a)u(b)(-1)(-1)x(i)cr(a,b) \end{aligned}$$

where $cr(z(i),a)$ = `cr(i,21)`, $cr(z(i),b)$ = `cr(i,22)`, and $cr(a,b)$ = `cr(21,22)`. The following MATLAB code segment may be written to evaluate $u(d(i))$.

```
for i=1:20
    ud(i) = sqrt( std(Z(i,:))^2 + (x(i)*std(B))^2 + std(A)^2 +
    2*std(Z(i,:))*std(B)*(-x(i))*cr(i,22) + 2*std(Z(i,:))*std(A)
    *(-1)*cr(i,21) + 2*std(A)*std(B)*(-x(i))*(-1)*cr(21,22) );
end
plot(ud)
```

25.8 Summary

A measurement result is incomplete without an accompanying statement of uncertainty. The GUM describes a method for computing the uncertainty in the output quantity given a functional relationship between the output quantity and the input quantities, and the standard uncertainties in the inputs. We briefly reviewed this method and illustrated its application in some simple filtering and fitting applications. The method however has limitations. In several instances, the relationship between the inputs and the output can only be expressed as a computer program. Further, computation of the confidence interval requires knowledge of the distribution of the output, something the GUM cannot provide. The Monte Carlo method is a numerical simulation method that overcomes these problems. In this chapter, we demonstrated how the GUM and the MCS may be used in propagating uncertainties in some simple filtering and fitting situations.

Exercises

25.1. Consider a sine wave of amplitude 1 μm and wavelength 0.8 mm, generated with a spacing of 1 μm over a length of 8 mm. Let the uncertainty in the amplitude at each sampling point be given by a normal distribution of zero mean and 0.1 μm standard deviation. Propagate these uncertainties through a filtering operation and determine the uncertainty in *Ra* and the interval corresponding to 95% level of confidence. Use Gaussian filter 0.8 mm cutoff.

25.2. In Exercise 25.1, repeat the process for *Rt*. Why is the uncertainty in *Rt* larger than that for *Ra*?

25.3. Consider a roundness profile R with a sinusoid of 10 UPR wavelength and 1 μm amplitude. Assuming uncertainty in the amplitude at each sampling point be given by a normal distribution of zero mean and 0.1 μm standard deviation, compute the uncertainty in the radius and center coordinates of the LS circle using the law of propagation of uncertainty (recall that the radius $r = (1/n)\sum_{i=1}^{n} R(i)$, $a = (n/2)\sum_{i=1}^{n} \cos\theta(i)R(i)$, and $b = (n/2)\sum_{i=1}^{n} \sin\theta(i)R(i)$). Verify using MCS.

25.4. What is the uncertainty in the residual profile in the previous exercise? Determine using the law of propagation of uncertainty and verify using MCS. What is the uncertainty in the radial out of roundness?

References

Cox, M.G., Dainton, M.P. and Harris, P.M. 2001, *Uncertainty and Statistical Modelling, Best Practice Guide No. 6, Software Support for Metrology Programme*, National Physical Laboratory. Teddington, UK.

Draper, N.R. and Smith, H. 1998, *Applied Regression Analysis*, 3rd edn, John Wiley & Sons. New York, USA.

Hopp, T. 1993, 'Computational Metrology', *Manufacturing Review*, vol. 6, no. 4, pp. 295–304.

Kreyzig, E. 1999, *Advanced Engineering Mathematics*, 8th edn, John Wiley & Sons. New York, USA.

Krystek, M. 2001, 'Measurement uncertainty propagation in the case of filtering in roughness measurement', *Measurement Science and Technology*, vol. 12, pp. 63–67.

Chapter 26
Error Separation Techniques in Roundness Metrology

26.1 Introduction

Methods to compensate spindle errors in roundness measurements are the focus of this chapter. The measured radial data as reported by a roundness instrument is a sum of both the radial form error of the part and the spindle's radial errors. A spindle generally has random and systematic components to its radial error motion, both of which convolve with the part surface. The techniques described in this chapter aim to isolate and remove the systematic component of the spindle's radial error from the measured data. There is always some uncertainty associated with any measurement including those made for the specific purpose of error separation, and we will review how the uncertainty in the raw data propagates to affect the results of error separation. As a more general note, reversals and error separation methods have long been used in metrology; Evans *et al.* (1996) present an excellent overview.

26.2 Full Reversal

A simple way of separating part error from spindle error involves measuring the part in two positions of the probe and the part. This method, referred to as full reversal (Bryan *et al.*, 1967; Donaldson, 1972), is the preferred technique for accurately separating spindle error from part error. First, the part is positioned at an arbitrary angular position on the spindle and a roundness trace is acquired. Then, the probe and the part are rotated by 180°. A new roundness trace is then acquired. Figure 26.1 shows the two reversal positions. If $Sx(\theta)$ is the spindle error along the sensitive direction (probing direction, say x) and $R(\theta)$ is the part error, then, in position 1 (see Fig. 26.1a), the probe reads a measured signal $m_1(\theta)$ given by

$$m_1(\theta) = R(\theta) + Sx(\theta)\,. \tag{26.1}$$

Fig. 26.1 **a** Position 1. **b** Position 2. A *rectangular key* is shown in both the *part* and the *spindle* to indicate relative angular positions at the start of each measurement

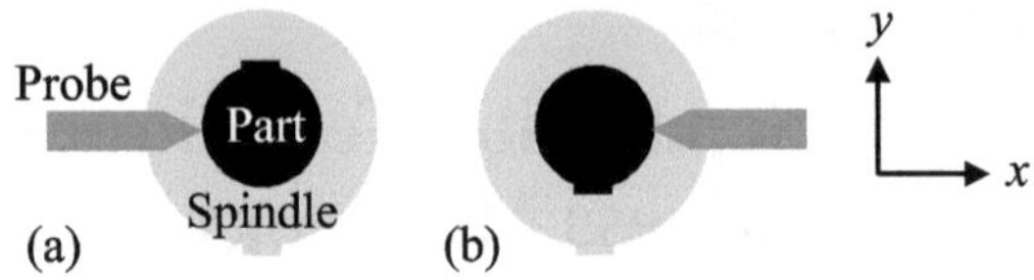

In position 2 (see Fig. 26.1b), the probe reads a measured signal $m_2(\theta)$ given by

$$m_2(\theta) = R(\theta) - Sx(\theta)\,. \tag{26.2}$$

Then, the part and spindle errors can be easily obtained as

$$\begin{aligned} R(\theta) &= (m_1(\theta) + m_2(\theta))/2 \\ Sx(\theta) &= (m_1(\theta) - m_2(\theta))/2\,. \end{aligned} \tag{26.3}$$

Note that the spindle will also have a systematic component to its radial error along the non-sensitive direction y. That component of the error does not convolve with part error in this measurement scheme, but will do in multi-probe methods.

26.3 Two-Position Method

While the full-reversal method accurately separates the spindle error from part error, it may not often be possible to rotate both the probe and the part by 180°. Therefore, there are other techniques that are typically used. These suffer from loss of accuracy but are simpler to implement. One such method is the two-position method (Whitehouse 1976), where only the part is rotated, not the probe. Figure 26.2 shows the part and probe orientations in the two positions. The measured signal at position 1 is

$$m_1(\theta) = R(\theta) + Sx(\theta)\,. \tag{26.4}$$

In position 2, the probe reads

$$m_2(\theta) = R(\theta - \pi) + Sx(\theta)\,. \tag{26.5}$$

The part and spindle errors can be separated by taking the Fourier transform of the probe outputs, subtracting them and subsequently taking the inverse Fourier transform. The Fourier transforms of the two equations are

$$mf_1(\omega) = Rf(\omega) + Sfx(\omega) \tag{26.6}$$

$$mf_2(\omega) = Rf(\omega)\,\mathrm{e}^{-j\omega\pi} + Sfx(\omega)\,. \tag{26.7}$$

The difference yields

$$Rf(\omega) = (mf_1(\omega) - mf_2(\omega))/(1 - \mathrm{e}^{-j\omega\pi})\,. \tag{26.8}$$

The inverse Fourier transform of Eq. 26.8 provides the part form error R.

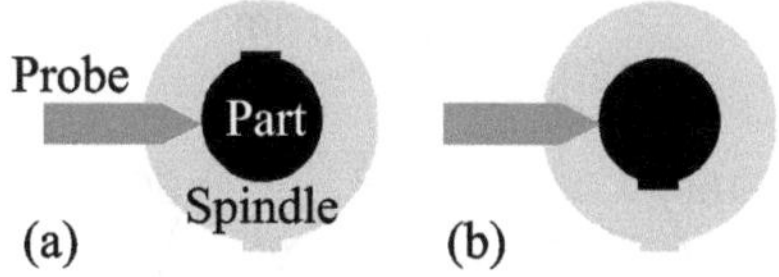

Fig. 26.2 **a** Position 1. **b** Position 2

The denominator in Eq. 26.8 is referred to as the transfer function, and is denoted by Hf. The problem with the above approach towards error separation is that Hf drops to zero for all even harmonics. That is, if the part were oval or had higher order even harmonics, this technique could not be used to reconstruct the part error accurately.

Example 26.1. Plot the transfer function Hf for different values of ω.

The transfer function can be computed and plotted as shown in Fig. 26.3. Notice that the transfer function is zero for all even harmonics.

```
n = 500;
omega=0:1:n/2+1;                    % Generate wavelength (UPR) axis
Hf = 1-exp(-j*omega*pi);            % Generate transfer function
plot(omega(1:50),abs(Hf(1:50))) % plot up to 50 UPR only
```

Example 26.2. Generate a roundness profile R containing 360 points with a 4 UPR harmonic of amplitude 1 µm. Assume that the systematic component of the spindle error, generally of higher frequency than that of the part, can be simulated by a 50 UPR signal of 0.02 µm amplitude. Generate profiles m_1 and m_2 as defined in Eqs. 26.4 and 26.5. Follow the analysis in Sect. 26.3 and separate the part profile from the spindle error. Is the reconstructed profile identical to the simulated profile R? If the part had 3 UPR signal instead of 4 UPR, does the reconstruction succeed?

The following code generates profiles m_1 and m_2. Notice that m_2 is generated by introducing a phase of 180° in the part profile R. There is no 180° phase in the

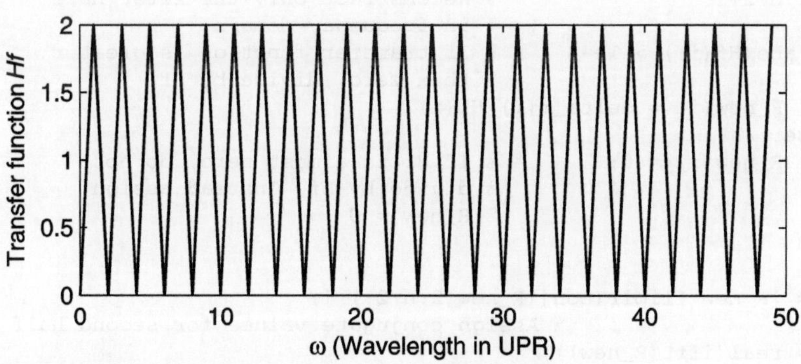

Fig. 26.3 Transfer function for the two-position method

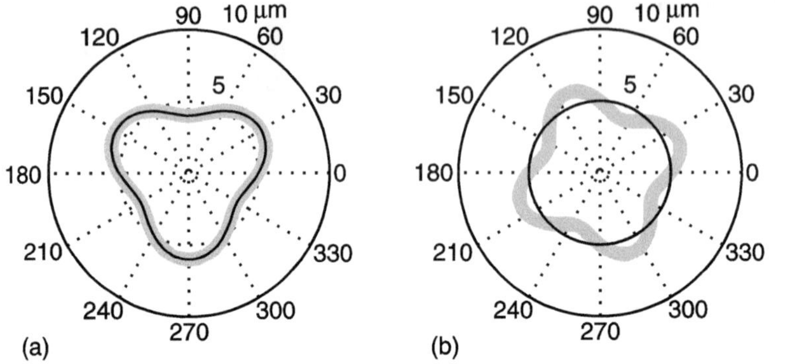

Fig. 26.4a,b Profile *R* (shown as *gray discrete points*) and the reconstructed profile (shown as a *black continuous line*) plotted on a nominal radius of 5 µm **a** for 3 UPR where the reconstruction has succeeded and **b** for 4 UPR where the reconstruction has failed

spindle error *Sx*. The reconstruction succeeds for 3 UPR but fails for 4 UPR part harmonic (see Fig. 26.4).

```
n = 360;
i = 0:1:n-1;
theta = i*2*pi/n;              % Generate theta
R = sin(2*pi.*i/n*4);          % Change this to
                               % R = sin(2*pi.*i/n*3);
                               % to generate a 3 UPR signal
Sx = 0.02*sin(2*pi.*i/n*50);  % Spindle error
m1 = R+Sx;                     % Profile m1
R1 = [R(n/2+1:n) R(1:n/2)];   % Introduce 180 degree phase
                               % in the part
m2 = R1+Sx;                    % Generate m2
omega=0:1:n/2;
Hf = 1-exp(-j*omega*pi);       % Transfer function
mf1 = fft(m1);                 % Take the Fourier transform
mf2 = fft(m2);
delta_m = mf1 - mf2;
for i=1:n/2+1                  % Reconstruct only the first half
                               % in frequency domain
    if abs(Hf(i)) > 1e-8       % If transfer function is greater
                               % than zero, divide by Hf
        R_new(i) = delta_m(i)/Hf(i);
    else
        R_new(i) = 0;          % If Hf is almost zero, do not
                               % divide by Hf. Instead assign
                               % R_new = 0
    end
end
R_new = [R_new fliplr(conj(R_new(2:n/2)))];
                          % Assign conjugate values for second half
R_new = real(ifft(R_new));
polar(theta,R+5,'k*'); hold on; polar(theta,R_new+5,'r-')
```

26.4 Two-Probe Method

A variation of the two-position method is the two-probe method. This method involves only one part position, but utilizes two probes (Whitehouse 1976) (see Fig. 26.5). Consider two probes placed 180° apart. The readings from the two probes are

$$m_1(\theta) = R(\theta) + Sx(\theta) \tag{26.9}$$
$$m_2(\theta) = R(\theta - \pi) - Sx(\theta)\,. \tag{26.10}$$

The part and spindle errors can be obtained by taking the Fourier transform of the probe outputs, adding them, and subsequently taking the inverse Fourier transform. This yields the following expression for $Rf(\omega)$:

$$Rf(\omega) = (mf_1(\omega) + mf_2(\omega))/(1 + \mathrm{e}^{-j\omega\pi})\,. \tag{26.11}$$

We leave it as an exercise to the reader to show that the transfer function $Hf = 1 + \mathrm{e}^{-jw\pi}$ drops to zero for all odd harmonics. That is, if a part were three-lobed or had higher order odd harmonics, this technique could not be accurately used to reconstruct the part error.

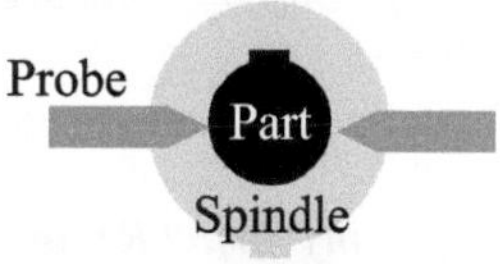

Fig. 26.5 Two-probe method

26.5 Three-Probe FFT Method

Both the two-position and two-probe methods suffer from harmonic suppression in that either all even or all odd harmonics remained undetected. To capture more harmonics, additional probes or more positions may be added. Several papers discuss the addition of multiple probes for this purpose (Aoki and Ozono 1966; Ozono 1974; Zhang *et al.*, 1997).

Consider the simple extension of the two-probe method to three probes, see Fig. 26.6. Let the three probes be located at 0°, ϕ and 2ϕ. It is clear the three probes see the same part signal, but shifted in phase. In addition to the part form error, the probes also see components of the spindle's radial errors. In the preceding sections, we only considered the spindle radial error motion along the sensitive direction, which was along the direction that the probes were aligned. At any given angular position, the spindle may also have error motion in an orthogonal direction, which would not have been detected by the probes in the arrangements discussed earlier but will play a role here.

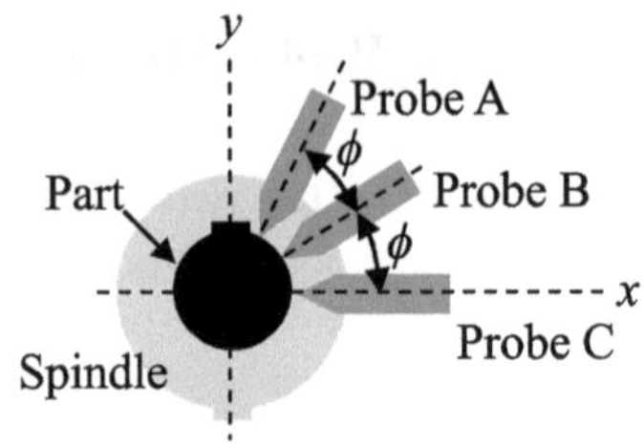

Fig. 26.6 Three-probe method

Let $R(\theta)$ represent the part form error, m_A, m_B and m_C represent the probe outputs, ϕ represent the angle between the probes and Sx and Sy represent the x and y components of the spindle error. Then,

$$\begin{aligned} m_A(\theta) &= R(\theta+2\phi)+Sx(\theta)\cos(2\phi)+Sy(\theta)\sin(2\phi) \\ m_B(\theta) &= R(\theta+\phi)+Sx(\theta)\cos(\phi)+Sy(\theta)\sin(\phi) \\ m_C(\theta) &= R(\theta)+Sx(\theta)\,. \end{aligned} \tag{26.12}$$

The part form error $R(\theta)$ is assumed to be measured from the x axis in a counterclockwise direction. Then, a differential output,

$$m(\theta) = m_c(\theta)-2\cos(\phi)m_B(\theta)+m_A\,(\theta) = R(\theta)-2\cos(\phi)R(\theta+\phi)+R(\theta+2\phi)\,, \tag{26.13}$$

is defined to remove the errors. The Fourier transform of Eq. 26.13 yields

$$mf(\omega) = Rf(\omega)(1-2\cos(\phi)\,\mathrm{e}^{j\omega\phi}+\mathrm{e}^{2j\omega\phi}) = Rf(\omega)Hf(\omega)\,, \tag{26.14}$$

where $mf(\omega)$ is the Fourier transform of $m(\theta)$ and $Rf(\omega)$ is the Fourier transform of $R(\theta)$.The part profile can then be obtained by taking the inverse transform of $mf(\omega)/Hf(\omega)$. Notice that the result $R(\theta)$ corresponds to an ideal profile (with no spindle error) measured at C in terms of phase.

We will see in the following example that the transfer function does drop to zero for certain harmonics, indicating that the harmonic suppression problem remains. However, there is marked improvement from the previous two techniques in that the number of harmonics that remain undetected is greatly reduced.

Example 26.3. Plot the transfer function for the three-probe technique for two different probe angles, (a) 10° and (b) 23°.

The transfer function of the three-probe technique can be computed and plotted as shown in Fig. 26.7. If ϕ is the probe angle in degrees, notice that the transfer function Hf drops to zero for harmonics given by $360n/\phi \pm 1$, where n is any integer. For $\phi = 10°$, harmonic suppression occurs at 1°, 35°, 37°, 71°, 73°, *etc.* When $\phi = 23°$, harmonic suppression occurs at non-integer values of 1°, 14.65°, 16.65°, 30.3°, 32,3°, *etc.* The use of prime numbers for probe spacing may be advantageous when using integer sampling intervals. It should be cautioned however that when the transfer function values are close to zero, reconstruction has larger uncertainties.

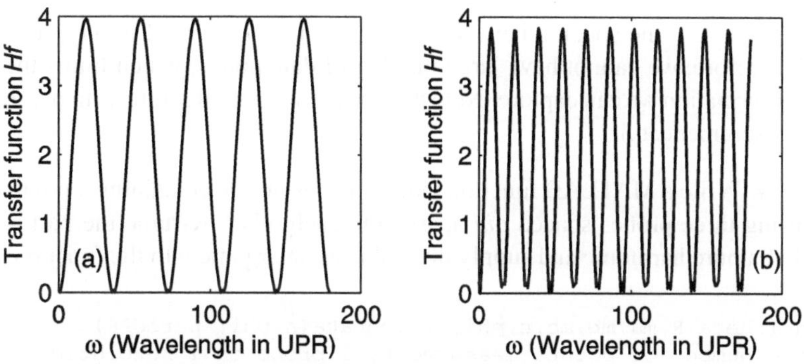

Fig. 26.7 Transfer function for probe angle of **a** 10° and **b** 23°

```
n = 360;
phi = 10*2*pi/360;                                % Probe angle of 10°
omega = 0:1:n/2+1;
Hf = 1 - 2*cos(phi)*exp(j*omega*phi) + exp(j*2*omega*phi);
                                                  % Transfer function
plot(omega(1:180),abs(Hf(1:180)))
phi = 23*2*pi/360;                                % Probe angle of 23°
omega = 0:1:n/2+1;
Hf = 1 - 2*cos(phi)*exp(j*omega*phi) + exp(j*2*omega*phi);
                                                  % Transfer function
plot(omega(1:180),abs(Hf(1:180)))
```

26.6 Three-Probe Sequential Method

The three-probe sequential method (Obi *et al.*, 1988; Obi and Kobayashi 1992; Gao and Kiyono 1996, 1997) is similar to the three-probe fast Fourier transform (FFT) method in that we use three probes to acquire data that is later used to separate the spindle error from part form error. In fact, the defining equations are identical to that in Sect. 26.5. The method is different from the FFT method in that we use an iterative reconstruction approach as opposed to the FFT approach.

As before, we have to remove $R(\theta)$ from $m(\theta)$; see Eq. 26.13. For this, we make a key observation. If the spacing between consecutive data points is equal to the probe angle, and if we know the first two points in the roundness profile, we can reconstruct the entire profile by rewriting the above equation in the following iterative form:

$$R(\theta+2\phi) = m(\theta) + 2\cos(\phi)R(\theta+\phi) - R(\theta)\,. \tag{26.15}$$

Because we don't know the first two points, we assume they are zero and proceed with the iterative reconstruction. Then, the result will be an eccentric circle that can be corrected by fitting and removing the best-fit limaçon (Muralikrishnan *et al.*, 2005). The result will be an accurate reconstruction of the profile. The disadvantage of this method is that the spacing between the points has to be equal to the probe angle and therefore, we cannot have large number of data points, which limits the amount of information we can capture. Note that the reconstructed profile is in phase with the measurement at *A*.

Example 26.4. Write a MATLAB function that simulates profile containing *n* points acquired using three probes spaced *phi* apart from each other. Assume the part is composed of a pure harmonic and supply this value as an argument to the function.

```
function [theta,R,ma,mb,mc,n,phi] = genData(n,phi, partUPR)
% n is the number of points measured by 1 probe in 1 revolution
% phi is the angle between probes in degrees
% simulating part error
theta = (0:1:n-1)'*2*pi/n;
R = sin(theta*partUPR);
R = R - mean(R);
% simulating spindle error (here we simulate spindle error as
% a large UPR term. Sx and Sy are components of the spindle's
% systematic radial error).
k = (1:n)';
Sx = 0.02*sin(2*pi.*k/n*50); Sx = Sx - mean(Sx);
Sy = 0.03*sin(2*pi.*k/n*40); Sy = Sy - mean(Sy);
% angle between probes
s = 360/n;    % s is the sampling interval in degrees
q = phi/s;    % the probe is located at every qth point
phi = phi*2*pi/360;
% the data that the probes see
mc = R + Sx; % Note that part profile R corresponds to mc
             % in terms of phase
R1 = [R(q+1:n); R(1:q)];
mb = R1 + Sx*cos(phi) + Sy*sin(phi);
R2 = [R(2*q+1:n); R(1:2*q)];
ma = R2 + Sx*cos(2*phi) + Sy*sin(2*phi);
```

Example 26.5. Use the above function and generate three probe data for $phi = 1°$, $n = 360$ and five lobed part form error. Then, reconstruct the part error using the three probe sequential method.

```
[theta,R,ma,mb,mc,n,phi] = genData(360,1,5);
phi =1*pi/180;
Ra(1,1) = 0;     % reconstructed profile at A is Ra
Ra(2,1) = 0;
for i = 3:360
    Ra(i,1) = (ma(i) -2*cos(phi)*mb(i) + mc(i))
              + 2*cos(phi)*Ra(i-1) -Ra(i-2);
end
polar(theta,R+4,'k*'); hold on; polar(theta,Ra+4,'r-')
```

Notice that the reconstructed profile is eccentric (Fig. 26.8). We therefore compute the best-fit limaçon and then obtain the deviations from this limaçon as shown below:

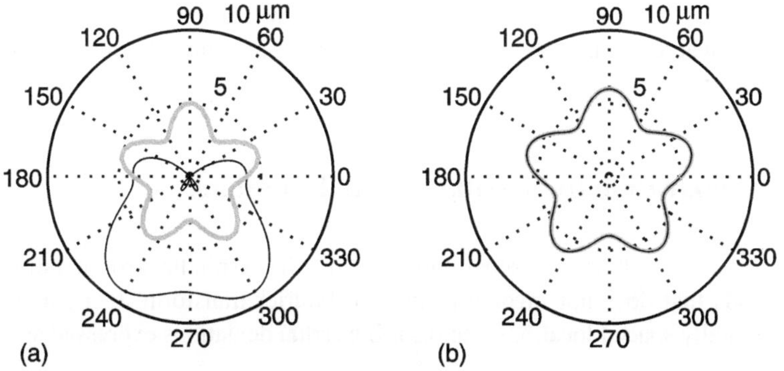

Fig. 26.8 **a** Reconstructed profile (*thin black line*) before removing the best-fit limaçon (cross-over at the *origin* indicates that the plot radius of 4 µm is not sufficiently large), and **b** after removing the best-fit limaçon. Original profile R is shown as *thick gray line*

```
n = length(Ra);
r = sum(Ra)/n;
a = 2*sum(Ra.*cos(theta))/n;
b = 2*sum(Ra.*sin(theta))/n;
r1 = r + a*cos(theta)+b*sin(theta);
Ra = Ra-r1;
% Rc is the reconstructed profile at C
Rc = [Ra(n-1:n); Ra(1:n-2)];
polar(theta,R+5,'k*'); hold on; polar(theta,Rc+5,'r-')
```

26.7 Three-Position FFT Method

The three-position FFT method utilizes three positions of the part to separate spindle error from part error, but with just one probe. The reconstruction is achieved using the Fourier and inverse Fourier transform. Let $R(\theta)$ be the part roundness error and $Sx(\theta)$ be the spindle error. Then

$$\begin{aligned} m_1(\theta) &= R(\theta) + Sx(\theta) \\ m_2(\theta) &= R(\theta+\phi_1) + Sx(\theta)\ , \\ m_3(\theta) &= R(\theta+\phi_2) + Sx(\theta) \end{aligned} \qquad (26.16)$$

where $0°$, ϕ_1, and ϕ_2 are the angles of the three positions. We can separate the errors by obtaining a new combined signal $m(\theta)$:

$$m(\theta) = 2m_1(\theta) - m_2(\theta) - m_3(\theta) = 2R(\theta) - R(\theta+\phi_1) - R(\theta+\phi_2)\ . \qquad (26.17)$$

As before, we take the FFT of the equation above to obtain

$$mf(\omega) = Rf(\omega)(2 - e^{j\omega\phi_1} - e^{j\omega\phi_2}) = Rf(\omega)Hf(\omega)\ . \qquad (26.18)$$

The magnitude of Hf drops to zero for those harmonics which satisfy $\omega = 360n_1/\phi_1 = 360n_2/\phi_2$, where n_1 and n_2 are any integers, thus reducing the number of harmonics that are suppressed.

26.8 Error Separation by Solving Linear Equations

There is an alternative technique of separating part error from spindle error in multi-position methods that does not involve taking the Fourier transform. If a part is sampled at n equally spaced locations, the resulting radial deviations expressed as

$$\begin{aligned} m_1(1) &= R(1) + Sx(1) \\ m_1(2) &= R(2) + Sx(2) \\ &\ldots \\ m_1(n) &= R(n) + Sx(n) \end{aligned} \tag{26.19}$$

form a sequence of n equations in $2n$ unknown. If the part is now rotated by some amount, say an amount corresponding to q sampling points, then we have the following set of n equations:

$$\begin{aligned} m_2(1) &= R(1+q) + Sx(1) \\ m_2(2) &= R(2+q) + Sx(2) \\ &\ldots \\ m_2(n-q) &= R(n) + Sx(n-q) \\ m_2(n-q+1) &= R(1) + Sx(n-q+1) \\ &\ldots \\ m_2(n) &= R(q) + Sx(n) \end{aligned} \quad . \tag{26.20}$$

These equations can be expressed in matrix form as $M = AX$, where M is a column vector of measured data, X is a column vector of unknowns and A is a matrix of 0 s and 1 s.

Although there are $2n$ equations (in $2n$ unknowns), they may not all be independent. Changing the spacing between the two part positions (q) will improve the rank of the matrix A (i.e., reduce the number of additional independent equations needed). Even so, an additional constraint such as setting the average spindle error equal to zero may be required to ensure full rank for A (enough independent equations to solve for $2n$ unknowns). This method also suffers from harmonic suppression, but is simpler to implement.

A modification of this method that avoids the need to compute the inverse of the matrix A involves measuring the part at p positions, where the angle between orientations is given by $360/p$. Under this special condition, the part form and spindle error can be separated by simply averaging selected rows. We illustrate this scheme for a simple case of measuring four sampling points and four positions. As before, let m_1, m_2, m_3, and m_4 be the four arrays containing four sampling points each, from profile measurement at four positions. Then,

$$\begin{array}{lcl} m_1(1) = R(1) + Sx(1) & & m_3(1) = R(3) + Sx(1) \\ m_1(2) = R(2) + Sx(2) & \text{and} & m_3(2) = R(4) + Sx(2) \\ m_1(3) = R(3) + Sx(3) & & m_3(3) = R(1) + Sx(3) \\ m_1(4) = R(4) + Sx(4) & & m_3(4) = R(2) + Sx(4) \\ \\ m_2(1) = R(2) + Sx(1) & & m_4(1) = R(4) + Sx(1) \\ m_2(2) = R(3) + Sx(2) & \text{and} & m_4(2) = R(1) + Sx(2) \\ m_2(3) = R(4) + Sx(3) & & m_4(3) = R(2) + Sx(3) \\ m_2(4) = R(1) + Sx(4) & & m_4(4) = R(3) + Sx(4) \end{array} . \qquad (26.21)$$

Averaging $m_1(1)$, $m_2(1)$, $m_3(1)$, and $m_4(1)$ yields the spindle error $Sx(1)$, offset by the average form error of the part (which is zero). Similarly, averaging $m_1(2)$, $m_2(2)$, $m_3(2)$, and $m_4(2)$ yields the spindle error $Sx(2)$, offset by the average form error of the part. Therefore, averaging every ith point of the measured profiles yields the ith point of the spindle error Sx. Similarly, we can select appropriate rows, which, upon averaging, yield the part form error. For example, averaging $m_1(1)$, $m_2(4)$, $m_3(3)$, and $m_4(2)$ yields the part form error $R(1)$ offset by the average spindle error.

In the example above, the number of positions chosen was equal to the number of sampling points. Under this condition, there will be no harmonic suppression. But the technique is not practical in that to measure 360 sampling points would require the part be positioned at 360 orientations, a task requiring huge time and effort.

A trade-off here is to reduce the number of positions while accepting harmonic suppression and the resulting loss of accuracy. In the preceding example, consider four sampling points and two positions. Therefore, in Eq. 26.21, we ignore m_2 and m_4 with the understanding that we have only measured profiles m_1 and m_3. Averaging $m_1(1)$ and $m_3(1)$ gives $Sx(1)$ offset by $(R(1)+R(3))/2$. On the other hand, averaging $m_1(2)$ and $m_3(2)$ gives $Sx(2)$ offset by $(R(2)+R(4))/2$. It can be shown through Fourier analysis (see (Estler *et al.*, 1997)) that these offsets are non-zero and more importantly, different, only for certain harmonics but zero for others. In this particular example, these offsets are zero for all odd harmonics and therefore the part form error can be reconstructed if the part contains odd harmonics only. The reconstruction is not accurate when the part contains even harmonics.

The technique is discussed in (Whitehouse 1976; Spragg 1967; Iizuka and Goto 1974; Reeve 1979). Multi-position method is studied in detail by Chetwynd and Siddall (1976) and Estler *et al.* (1997) discuss uncertainty aspects of the technique.

26.9 Uncertainty Propagation

The techniques we discussed in this section pertain to the removal of the systematic component of a spindle's radial error from the part form error. In the discussions thus far, we did not consider uncertainty in the measured data. We take a brief look at how uncertainty propagates through the error separation process.

Consider full reversal discussed in Sect. 26.2. Let the uncertainty in each measured point of profiles m_1 and m_2 in Eq. 26.3 be given by $u(m(\theta))$. Then, from the law of propagation of uncertainty, the uncertainty in part form R is given by

$$u(R(\theta)) = \frac{1}{2}\sqrt{u^2(m(\theta)) + u^2(m(\theta))} = \frac{u(m(\theta))}{\sqrt{2}}. \qquad (26.22)$$

This result is intuitive in that the part error is the average of two measurements and averaging results in reduced uncertainty. Estler *et al.* (1997) showed that the uncertainty in a p-position multi-position method (p equally spaced positions that span the entire circle) is given by

$$u(R(\theta)) = \frac{u(m(\theta))}{\sqrt{p}}, \qquad (26.23)$$

where $u(m(\theta))$ is the uncertainty in each measured point at each of the p positions. MCS may also be used to verify the above results.

26.10 Summary

Error separation is an important aspect of precision roundness metrology. Precision spindles have undesired radial motion that is superimposed on the part form error. A portion of the spindle's radial error motion is systematic and can therefore be mapped and removed. For this purpose, a high precision ball is generally mounted on the spindle and measured using multiple probes and/or multiple positions of the ball on the spindle. Mathematical techniques can then be applied to the data to separate spindle error from part error. We studied several such error separation methods in this chapter. Further, there is always some uncertainty associated with every measurement and this uncertainty propagates through the error separation process. We briefly reviewed uncertainty propagation in the context of full reversal.

Exercises

26.1. For the data in Example 26.2, reconstruct the profile using the two-probe method. Verify that the reconstruction is perfect for even harmonics, but fails for odd harmonics.

26.2. Generate three-probe data using `genData` function (see Example 26.4) with $n = 360$ points, and probe angle of 10°, and a part containing a pure 3rd harmonic of amplitude 1 μm. Reconstruct the profile using the three-probe FFT method. What happens when the probe angle is still 10° but the part has a pure 35th harmonic of 1 μm amplitude?

26.3. If the probe angle were changed to 23°, does the reconstruction succeed for the above problem for the two cases? Why?

26.4. Generate three-probe data using `genData` function (see Example 26.4) with $n = 360$ points, and probe angle of 1°, and a part containing a pure 23rd harmonic of amplitude 1 μm. Reconstruct the profile using the three-probe sequential method. Is there ever a harmonic suppression in this method? If not, what limits the use of this technique for any and all applications?

26.5. Plot the transfer function for the three-position technique assuming $n = 360$ and the part is rotated by 10° between positions (sampling interval is 1°), and verify the claim that the zeros occur as stated in Sect. 26.7. Based on this, suggest a method to reduce the number of harmonics that are suppressed.

26.6. Write a function to generate three-position data in a similar manner to that in Example 26.4.

26.7. Generate three-position data for $n = 360$ points, and position angle of 10°, and a part containing a pure 3rd harmonic of amplitude 1 μm. Reconstruct the part form error (a) using the three position FFT method, and (b) as suggested in Sect. 26.8 by solving a set of linear equations. Explain the results.

26.8. Consider a part containing a pure 3rd harmonic of amplitude 1 μm (let $n =$ 360). Assume the systematic component of the spindle error may be simulated using a random number generator with amplitude between 0 and 0.3 μm. First synthesize the observed readings of the probe at the two positions of a full reversal. Assume that the standard uncertainty in each point of measured profile at both positions of the reversal is 0.1 μm. What is the uncertainty in the reconstructed part profile? Determine using both the GUM and MCS methods.

References

Aoki, Y. and Ozono, S. 1966, 'On a new method of roundness measurement based on the three-point method', *Journal of the Japan Society for Precision Engineering*, 32, no. 12, pp. 831-836.

Bryan, J., Clouser, R. and Holland, E. 1967, 'Spindle accuracy', *American Machinist*, vol. 612, pp. 149–64.

Chetwynd, D.G. and Siddall, G.J. 1976, 'Improving the accuracy of roundness measurement', *Journal of Physics E: Scientific Instruments*, vol. 9, pp. 537–544.

Donaldson, R.R. 1972, 'A simple method for separating spindle error from test ball roundness error', *Annals of the CIRP*, vol. 21, pp. 1, pp.125–126.

Estler, W.T., Evans, C.J. and Shao, L.Z. 1997, 'Uncertainty estimation for multiposition form error metrology', *Precision Engineering*, vol. 21, no. 2, pp. 72–82.

Evans, C.J., Hocken, R.J. and Estler, W.T. 1996, 'Self-calibration: reversal, redundancy, error separation, and 'absolute testing'', *Annals of the CIRP*, vol. 45, no. 2, pp. 617–634.

Gao, W. and Kiyono, S. 1996, 'High accuracy profile measurement of a machined surface by the combined method', *Measurement*, vol. 19, no. 1, pp. 55–64.

Gao, W. and Kiyono, S. 1997, 'On-machine roundness measurement of cylindrical workpieces by the combined three-point method', *Measurement*, vol. 21, no. 4, pp. 147–156.

Iizuka, K. and Goto, M. 1974, 'A new precision measuring system of roundness and cylindricity with self calibrating capability', *Proceedings of the International Conference on Production Engineering*, Tokyo, pp. 451–456.

Muralikrishnan, B., Venkatachalam, S., Raja, J. and Malburg, M. 2005, 'A note on the three-point method for roundness measurement', *Precision Engineering*, vol. 29, no. 2, pp. 257–260.

Obi, M., Kobayashi, T. and Furukawa, S. 1988, 'On a new method of roundness measurement based on the sequential-three-points method', *Transactions of the Japan Society of Mechanical Engineers. C*, vol. 54, no. 506, pp. 2475–2480.

Obi, M. and Kobayashi, Y. 1992, 'A measuring method of the displacement along a circle based on the sequential-three-point method (A study of the measuring principle and error)', *Transactions of the Japan Society of Mechanical Engineers. C*, vol. 58, no. 548, pp. 1278–1283.

Ozono, S. 1974, 'On a new method of roundness measurement based on the three points method', *Proceedings of the International Conference on Production Engineering*, Tokyo, pp 457–462.

Reeve, C. 1979, *Calibration of a Roundness Standard*, National Bureau of Standards (National Institute of Standards and Technology) Internal Report, NBSIR, 79-1758. Gaithersburg, MD, USA.

Spragg, R.C. 1967, 'Accurate calibration of surface texture and roundness measuring instruments', *Proceedings of the Institution of Mechanical Engineers*, vol. 182, no. 3K, pp. 397–405.

Whitehouse, D.J. 1976, 'Some theoretical aspects of error separation techniques in surface metrology', *Journal of Physics E: Scientific Instruments*, vol. 9, pp. 531–536.

Zhang, G.X., Zhang, Y.H., Yang, S.M. and Li, Z. 1997, 'A multi-point method for spindle error motion measurement', *Annals of the CIRP*, vol. 46, no. 1, pp. 441–445.

Chapter 27
Other Relevant Topics

27.1 Introduction

Fitting, filtering, parameterization and uncertainty determination are at the heart of any surface texture or roundness analysis. We have presented these topics from an implementation standpoint that will hopefully be of value to students and practitioners alike. The topics we have covered include methods very commonly used in industry, and some other emerging ones. We should point out that surface and roundness metrology is an active research area and new computational methods are constantly introduced. We present here a brief abstract of some areas of research in computational surface and roundness metrology.

27.2 Parameter Extraction

The determination of functionally relevant parameters is a critical aspect of surface texture analysis (De Chiffre *et al.*, 2000). The central premise in filtering surface profiles or area maps is that different surface-wavelength bands affect functionality in different ways. A parameter computed on the entire profile may not encode the functionality, while the same parameter computed on a narrow bandwidth may clearly do so. Such surface-wavelength-based filtering and parameterization is generally accepted and practiced in industry but is not the only approach to identifying functionally relevant parameters. In fact, we discussed a rolling-ball-based filtering technique where the profile or area map may be partitioned into finer bands on the basis of feature sizes. Parameters may be computed on such partitioning of the profile or area map.

The French automotive industry has been utilizing a method, gradually refined over several decades, that is different yet effective. Their method is referred to as the motif method (International Organization for Standardization 1996). A motif is a small segment of the profile containing two peaks. The analysis procedure involves identifying the different motifs and later deciding if any two adjacent motifs may be combined because one of them is insignificant. The final set of motifs provide different parameters such as average motif depth, average width between peaks, *etc.* More recently, the method has been extended to 3D area maps where hills and dales (Scott 1998) are identified and combined to provide qualitative and quantitative assessments of surface texture.

Another approach to surface texture analysis is based on the idea that engineering surfaces demonstrate fractal behavior (Sayles and Thomas 1978; Thomas and Thomas 1988; Russ 1994; Brown *et al.*, 1993). That is, the structure repeats itself at different scales, and the level at which the surface exhibits such self-similarity yields parameters that may be related to functional performance. The ASME B46.1 (American Society of Mechanical Engineers 2002) incorporates fractal analysis in its Standard.

27.3 Tools for Correlation and Diagnostics

The ultimate objective of any surface texture analysis, however, is not in the determination of parameters themselves, but rather in capturing the relationship between these parameters (say *Ra*, *Rq*, *etc*) and measures of functional performance of the component (coefficient of friction, optical reflectivity, sealing capability, *etc*). In the diagnostic mode, it is also the objective of surface texture analysis to elucidate any relationship between surface texture parameters and manufacturing process parameters (such as tool wear, feed rate, depth of cut, chatter, *etc*). Establishing such relationships enables the prediction of functional performance of a component or in diagnosing manufacturing process variability; a key contribution in terms of value addition in metrology.

There are several tools that may be used for this purpose. Assuming quantitative functional performance data is available along with surface texture parameters, statistical regression may be used to establish the relationship between the two. For example, suppose that surface texture and optical reflectivity of 100 polished surfaces are measured. Regression models may be used to determine which surface texture parameters to incorporate into the model, and the exact nature of the dependence of reflectivity on those parameters can then be determined.

In recent times, artificial neural network (ANN)-based systems have been incorporated for this purpose (Mainsah and Ndumu 1998; Muralikrishnan and Raja 2003). With the availability of training data, the ANNs may be trained to perform a variety of tasks, such as discriminate between parts from different manufacturing processes, differentiate between parts from the same manufacturing process but manufactured under different conditions (say, to identify the progression of tool wear), distinguish parts based on binary or multi-valued functional performance information, *etc.*

In addition, where no functional performance or manufacturing process information is available, ANN-based systems may be used in pure clustering mode to partition the set of components into groups based on parameter values. A sample from each group may then be tested to assess if they possess significantly different performance from the other.

In general, real applications involve a multitude of effects and therefore their complexity precludes experimental determination of appropriate bounds for different surface texture parameters. It is general practice in industry, therefore, to specify tolerances of surface texture parameters based on experience. A part that meets the tolerance is more likely to satisfy its performance requirements than one that does not. The relationship between functional performance and surface texture is therefore based on simple rules such as:

If $Ra < 2.3\,\mu\text{m}$ and $Sm < 100\,\mu\text{m}$, component satisfies functional performance
Else it does not.

Such rule-based systems have abrupt transitions at the boundaries (what happens to two parts, one with $Ra = 2.29\,\mu\text{m}$ and another with $Ra = 2.31\,\mu\text{m}$?). These rule-based systems may be modified using fuzzy logic that incorporates user experience effectively.

27.4 Challenges

3D characterization of surface texture is now increasingly used (Lonardo *et al.*, 1996; Blunt and Jiang 2003) instead of 2D profile measurement and analysis. Characterization of 3D surfaces is an active research topic. With the collection of 3D data, there is distinct possibility of demonstrating relationship between manufacturing process parameters, surface texture and a component's functionality in a large variety of applications. Efforts in this direction are the major undertakings of several research groups around the world.

The development of software analysis tools and their validation is also another major challenge facing the industry. A number of national metrology institutes have provided reference data and/or reference software against which independent software vendors may validate their algorithms and implementations. As more advanced mathematical techniques are incorporated into such software systems, there will be a growing need to provide further reference algorithms and data sets.

The analysis of structured surfaces, those with repetitive patterns, will involve entirely new methods and techniques (Evans and Bryan 1999; Stout and Blunt 2001; Verma and Raja 2005). Surface-wavelength-based filtering and average parameters will be less applicable in such instances, while discrete feature geometry based parameters and their statistics over the surface appear to be more relevant.

References

American Society of Mechanical Engineers 2002, *ASME B46.1-2002, Surface Texture (Surface Roughness, Waviness and Lay)*, ASME. New York, NY, USA.

Blunt, L. and Jiang, X. (eds) 2003, *Advanced Techniques for Assessment Surface Topography: Development of a Basis for 3D Surface Texture Standards*, Kogan Page Science. London, UK.

Brown, C.A., Charles, P.D., Johnsen, W.A. and Chesters, S. 1993, 'Fractal analysis of topographic data by the patchwork method', *Wear*, vol. 161, nos. 1–2, pp. 61–67.

De Chiffre, L., Lonardo, P., Trumpold, H., Lucca, D.A., Goch, G., Brown, C.A., Raja, J. and Hansen, H.N. 2000, 'Quantitative characterization of surface texture', *Annals of the CIRP*, vol. 49, no. 2, pp. 635–652.

Evans, C.J. and Bryan, J.B. 1999, 'Structured, textured or engineered surfaces', *Annals of the CIRP*, vol. 48, no. 2, pp. 541–556.

International Organization for Standardization 1996, *ISO 12085:1996, Geometrical Product Specifications (GPS) – Surface Texture: Profile Method – Motif Parameters*, ISO. Geneva, Switzerland.

Lonardo, P.M., Trumpold, H. and De Chiffre, L. 1996, 'Progress in 3D surface microtopography characterization', *Annals of the CIRP*, vol. 45, no. 2, pp. 589–597.

Mainsah, E. and Ndumu, D.T. 1998, 'Neural network applications in surface topography', *International Journal of Machine Tools and Manufacture*, vol. 38, nos. 5–6, pp. 591–598.

Muralikrishnan, B. and Raja, J. 2004, 'Process diagnostics and functional correlation in surface metrology: novel techniques, case studies and analysis system development', *Measurement*, vol. 36, no. 2, pp. 175–183.

Russ, J.C. 1994, *Fractal Surfaces*, Plenum Press. New York, NY, USA.

Sayles, R. and Thomas, T.R. 1978, 'Surface topography as a non-stationary random process', *Nature*, vol. 271, pp. 431–434.

Scott, P.J. 1998, 'Foundations of topological characterization of surface texture', *International Journal of Machine Tools and Manufacture*, vol. 38, nos. 5–6, pp. 559–566.

Stout, K.J. and Blunt, L. 2001, 'A contribution to the debate on surface classifications – random, systematic, unstructured, structured and engineered', *International Journal of Machine Tools and Manufacture*, vol. 41, nos. 13–14, pp. 2039–2044.

Thomas, T.R. and Thomas, A.P. 1988, 'Fractals and engineering surface roughness', *Surface Topography*, vol. 1, pp. 143 152.

Verma, R. and Raja, J. 2005, 'Characterization of engineered surfaces', *Journal of Physics: Conference Series*, vol. 13, pp. 5–8.

Answers to Selected Exercises

3.1 zf, the FFT of array z, is $[12 \quad -2+4i \quad -4 \quad -2-4i]$.

3.2 The 5th and the 17th terms of zf (the FFT of z) contain the required information.

3.3 See Example 5.2. The attenuation at the cutoff is 50%. The filter has linear phase. Note that when computing the phase, the array might have values of 0, 2π or -2π, all of which are equal.

3.4, 3.5 See Example 6.1.

4.1 We obtain distortion-free data only when the window is completely within the profile. This condition is satisfied for points on the profile between $(m-1)/2+1$ to $n-(m-1)/2$, when m is odd.

4.2 Implementing convolution in frequency domain is computationally advantageous.

4.3 The amplitude of the output sinusoid is 0.0012 µm with a wavelength of 0.1 mm.

4.5 The amplitude of the filtered profile is 0.3956 µm (discard data near the edges where the amplitude may be larger).

4.6 Yes, it does.

4.7 Yes, except at the edges.

4.8 The filtered profile matches that obtained in Exercise 4.5 over the entire 8 mm length.

5.4 The amplitudes of w_1 (waviness profile of z_1) and w_2 (waviness profile of z_2) are 0.895 µm and 0.085 µm respectively.

5.5 The waviness profile w (corresponding to $z = z_1 + z_2$) $= w_1 + w_2$, where w_1 is a sinusoid of amplitude 0.895 µm and wavelength 2 mm, and w_2 is a sinusoid of amplitude 0.085 µm and wavelength 0.5 mm.

5.8 *Wa* values are 0.6160 µm, 0.5771 µm, and 0.3182 µm. *Wa* decreases with increasing cutoff, because more of the surface wavelengths are now part of roughness.

6.2 The amplitude and phase of the first sinusoid after filtering are 0.3243 μm and 1.9298 rad. The amplitude and phase of the second sinusoid after filtering are 0.8848 μm and 0.6926 rad. The waviness profile w (corresponding to $z = z_1 + z_2$) = $0.3243\sin(2\pi x/2 - 1.9298) + 0.8848\sin(2\pi x/0.5 - 0.6926)$.

7.1 $Wa = 0.842$ μm (computed over the entire length of the profile).

7.3 The 2RC filter equation for roundness profiles is

$$S(i) = A\omega c\,(2 - Ai\omega c/n)\exp(-Ai\omega c/n)\ , \quad i = 1:n\,.$$

7.4 $Wa = 1.1091$ μm. Note, however, that we have used only an approximation to the waviness profile. The true 2RC waviness profile has to be determined using the roundness version of the equation in Exercise 6.1.

7.5 The 2RC transmits 75% at the cutoff unlike the Gaussian which only transmits 50%. Therefore, the filtered roughness profile shows larger amplitudes for the 2RC.

7.6 For the 10 mm diameter part, $\lambda c = \frac{\pi D}{\omega c} = \frac{10\pi}{20} = 1.5708$ mm. For the 5 mm diameter part, $\lambda c = \frac{\pi D}{\omega c} = \frac{5\pi}{20} = 0.7854$ mm.

7.7 The plot radius should be larger than the smallest measured deviation, and should not be much larger than the largest measured deviation. A large value will not highlight the actual form, while a small value may result in zero crossings.

8.2 In Example 8.1, replace $z(j,i) = 2\sin(2\pi x(i)/6.4)$ by $z(j,i) = 2\sin(2\pi y(j)/6.4)$. Then follow the same procedure as in the examples. The resulting amplitude after filtering is 1 μm.

8.3 In Example 8.1, replace $z(j,i) = 2\sin(2\pi x(i)/6.4)$ by $z(j,i) = 2\sin(2\pi x(i)/6.4) + 2\sin(2\pi y(j)/6.4)$. Then follow the same procedure as in the examples.

8.4 Recall that the low-pass filter retains low frequencies or large wavelengths. Therefore, by choosing a cutoff wavelength smaller than the one we wish to retain, we can achieve the objective along x. Similarly, we choose a cutoff larger than the one we wish to discard along y. Therefore, we choose 0.8 mm for the x cutoff and 25.6 mm for the y cutoff.

8.5 Yes, they should match identically even at the edges.

9.1 The ratio of sums (at center to that at beginning) is 1.995. This is because the filter at the edge has half the area of that at the center. The zero-order regression corrects for this by scaling each filter point at the beginning of the profile to twice its nominal value. Therefore the filtered profile point has twice the nominal value also. Continuing from Example 9.2, $w_2(1)/w_1(1) = 1.995$.

9.2 In Example 9.2, we determine the filter window and perform the weighted sum for every point on the profile. This amounts to performing a time-domain convolution. Instead we can use MATLAB's `conv` function and determine the profile but use the method described in Example 9.2 only for the edge points (one cutoff length) on either side.

9.4 Differentiating with respect to the unknowns yields the following set of equations that can be simplified into matrix form $MP = Q$, where M is 6×6, and P and Q are 6×1.

$$\sum_{py=1}^{ny} \sum_{px=1}^{nx} (z(px,py) - Ax(kx,px)^2 - By(ky,py)^2 - Cx(kx,px) - Dy(ky,py) \\ - Ex(kx,px)y(ky,py) - F)S(kx,px,ky,py)x(kx,px)^2 = 0$$

$$\sum_{py=1}^{ny} \sum_{px=1}^{nx} (z(px,py) - Ax(kx,px)^2 - By(ky,py)^2 - Cx(kx,px) - Dy(ky,py) \\ - Ex(kx,px)y(ky,py) - F)S(kx,px,ky,py)y(ky,py)^2 = 0$$

$$\sum_{py=1}^{ny} \sum_{px=1}^{nx} (z(px,py) - Ax(kx,px)^2 - By(ky,py)^2 - Cx(kx,px) - Dy(ky,py) \\ - Ex(kx,px)y(ky,py) - F)S(kx,px,ky,py)x(kx,px) = 0$$

$$\sum_{py=1}^{ny} \sum_{px=1}^{nx} (z(px,py) - Ax(kx,px)^2 - By(ky,py)^2 - Cx(kx,px) - Dy(ky,py) \\ - Ex(kx,px)y(ky,py) - F)S(kx,px,ky,py)y(ky,py) = 0$$

$$\sum_{py=1}^{ny} \sum_{px=1}^{nx} (z(px,py) - Ax(kx,px)^2 - By(ky,py)^2 - Cx(kx,px) - Dy(ky,py) \\ - Ex(kx,px)y(ky,py) - F)S(kx,px,ky,py)x(kx,px)y(ky,py) = 0$$

$$\sum_{py=1}^{ny} \sum_{px=1}^{nx} (z(px,py) - Ax(kx,px)^2 - By(ky,py)^2 - Cx(kx,px) - Dy(ky,py) \\ - Ex(kx,px)y(ky,py) - F)S(kx,px,ky,py) = 0$$

10.2 The spline and the zero-order have excellent edge performance. The spline is computationally more efficient.

10.3 The Gaussian filter waviness profile appears to have reduced edge effects because the profile starts and ends at zero. The Gaussian filter always tends to draw the waviness profile towards the zero line near the edges (because the moving window sees fewer surface points). The spline filter, on the other hand, attempts to fit a spline to all points within the window. Near the edge, the spline relies on points to one side only. When the filter window is large or equivalent to the dominant surface wavelength, the fitting process is not a good representation of the mean line near the edge. On the other hand, when the filter window is smaller in comparison to the dominant surface wavelength, the spline produces an excellent fit, while the Gaussian shows some edge effect.

10.4 Because the profile length is not an integer multiple of 0.8 mm (wavelength of sinusoid), the Gaussian filter attempts to bring the mean line down to zero near the end while the spline attempts to keep up with the profile. When cutoff is reduced to 0.08 mm, the spline agrees with the profile remarkably well; the Gaussian shows distinct edge effects.

10.5 If part radius is 0.573 mm, sampling interval is $(2\pi)0.573/3{,}600 = 0.001$ mm, and profile length is 3.6 mm with 50 undulations in that length (cutoff is 50 UPR), cutoff is therefore $3.6/50 = 0.072$ mm. The resulting waviness profiles are almost identical. The periodic spline filter wraps the profile around and so performs a slightly better, but hardly noticeable, fit.

11.2 Twice the cutoff at the beginning and twice the cutoff at the end (assuming the filter is generated from −cutoff to +cutoff).

11.3 Increasing the cutoff provides a mean line more robust against outliers. That does not however imply that increasing the cutoff is the solution to overcome this problem. A cutoff is primarily decided on the basis of functional requirements.

12.1 An efficient implementation is provided below:

```
function C1 = f_dilate(z,B);
n = size(z,1);  % size of row vector z
m = size(B,1);  % size of row vector B
% first process the interior points.
for i = m:n - m + 1
    C1(i) = z(i) + B(1); %initialize
    for m = 2:m % iterate to obtain "convolution"
        C1(i) = max(C1(i), z(i+m-1)+B(m));
        C1(i) = max(C1(i), z(i-m+1)+B(m));
    end
end
% Next, process sets of the end points on both sides.
for j = 1: m-1
    C1(j) = z(j) + B(1);          %initialize the left end points
    C1(n-j+1) = z(n - j+1) + B(1);% initialize the
                                  % right end points
    for k = 1:j           % iterate to obtain end-half "convolution"
        C1(j) = max(C1(j), z(j-k+1)+ B(k));
        C1(n - j +1) = max(C1(n-j+1), z(n-j+k)+ B(k));
    end
    for k = 2:m                   % iterate to obtain the other
                                  % half "convolution"
        C1(j) = max(C1(j), z(j+k-1) + B(k));
        C1(n - j+1) = max(C1(n-j+1), z(n-j+2-k)+B(k));
    end
end
function B = ballel(radius, delta);
m = fix(radius/delta) ;
for m = 1:m %generate "first quadrant" ball points
    B(m,1) = sqrt(radius^2 - ((m - 1)*delta)^2);
end
```

12.3 C_3 will not be identical to C_1, but will be a close approximation to it. The idea of partitioning a profile into different scales – each generated from a structuring element of different size is advantageous in identifying scale dependence on profiles – such as outlier analysis. See Chap. 13.

12.4 The *closing* mean line produces an envelope that hugs the surface. Applications involving mating of conformable surfaces might conceivably be modeled using the *closing* mean line. Application involving rolling balls on rough surfaces can conceivably be modeled using the *dilation* mean line.

13.1 No, the difference and approximation profiles will not change.

13.2 It is not an ideal choice because it does not have steep transmission characteristics. A step filter implemented in the frequency domain will be a better choice for this purpose.

13.3 If a large structuring element is used in the first iteration, the difference profile contains all the details. This difference profile can then be used as the profile for the next step, where a smaller structuring element can be applied.

13.4 Level 8.

15.1 The slope of the LS line is 0.0655. The LS line is not only influenced by the amplitude of $z(i)$, but also by its location $x(i)$. Notice that there is an uneven distribution of the points along x. That is, the first 1 mm along x has points that lie above the axis, while the last 1 mm has points below.

15.2 Direction cosines of the plane $(a,b,c) = (0.707, -0.707, 0)$. The centroid is $(3.95, 3.95, 0.9875)$.

15.3 The centroid $(x_c, z_c) = (\frac{\sum_{i=1}^{n} x(i)}{n}, \frac{\sum_{i=1}^{n} z(i)}{n})$. Substitute for x as x_c in the equation for the line and show that $z = z_c$.

15.5 The function for best-fit plane is identical to that written for a 3D line in Example 15.4, except that that we identify the eigen vector corresponding to the smallest eigen value. Therefore, we replace the line `[y,i]= max(a1)` with the line `[y,i]= min(a1)`.

16.1 Steepest descent algorithm:

```
function [xc,zc,r] = LSCircleInPlaneSD(M,xco,zco,ro)
eps = 1e-8;
flag = 0;
lamda = 0.1;
n = size(M,1);
Eo = sum((sqrt((M(:,1)-xco).^2+(M(:,2)-zco).^2)-ro).^2);
% Steepest descent iteration for LS best-fit circle
t = 1;
while flag == 0
% Jacobian
    for i=1:n
        R = sqrt((M(i,1)-xco)^2+(M(i,2)-zco)^2);
        dexo(i)  = -2*(R-ro)*(M(i,1)-xco)/R;
        dezo(i)  = -2*(R-ro)*(M(i,2)-zco)/R;
        dero(i)  = -2*(R-ro);

    end
```

```
    xco = xco - lamda*sum(dexo);
    zco = zco - lamda*sum(dezo);
    ro = ro - lamda*sum(dero);
    E1 = sum((sqrt((M(:,1)-xco).^2+(M(:,2)-zco).^2)-ro).^2);
    if abs(Eo-E1)<eps, flag = 1; end;
    Eo = E1;
    disp(['iteration # ' num2str([t xco zco ro Eo])]);
    t = t + 1;
end
xc = xco; zc = zco; r = ro;
```

16.2 Levenberg–Marquardt algorithm:

```
function [xc,zc,r] = LSCircleInPlaneLM(M,xco,zco,ro)
eps = 1e-8;
flag1 = 0;
n = size(M,1);
% Levenberg-Marquardt Iteration for LS best-fit circle
lamda = 0.0001;
it1 = 1;
while flag1 == 0
    flag2 = 0;
    it2 = 1;
    % Jacobian
    for i=1:n
        d(i,1) = sqrt((M(i,1)-xco)^2+(M(i,2)-zco)^2) - ro;
        Fo(i,1) = -(M(i,1)-xco)/(d(i,1)+ro);
        Fo(i,2)=-(M(i,2)-zco)/(d(i,1)+ro);Fo(i,3)=-1;
    end
    U = Fo'*Fo;
    v = Fo'*d;
    Eo = (sum(d.*d));
    while flag2 == 0
        H = U + lamda*(eye(size(U))+eye(size(U)).*U);
        x = H\(-v);
        xc1 = xco + x(1);
        zc1 = zco + x(2);
        r1 = ro + x(3);
        for i=1:n
            d(i,1) = sqrt((M(i,1)-xc1)^2+(M(i,2)-zc1)^2) - r1;
        end
        E1 = (sum(d.*d));
        if E1 < Eo
            flag2 = 1; xco = xc1; zco = zc1; ro = r1;
            lamda = lamda * 0.04;
        else
            lamda = lamda * 10;
        end
        it2 = it2 + 1;
    end
    if abs(Eo - E1) < eps
        flag1 = 1;
    end
    disp(['iteration # ' num2str([it1 xco zco ro E1])]);
```

```
    it1 = it1 + 1;
end
xc = xco; zc = zco; r = ro;
```

16.3 The center and radius are $(0.2612, 0.3433)$ and 10.0216.

16.4 Increasing λ increases the speed of convergence, while decreasing λ minimizes the risk of overshooting the minima.

16.5 The steepest descent algorithm generally takes a few more iterations to converge.

16.6 To use MATLAB's `lsqnonlin` function, we first write a function `MatlabNonLinearLSCircle`, which we subsequently call as shown below. Note that the function `MatlabNonLinearLSCircle` is required to output deviations at each point, not the sum of squared deviations. x is an array of size 3×1 containing the current estimates of the center and radius, and M is a $n \times 2$ array containing the data points.

```
function F = MatlabNonLinearLSCircle(x,M)
for i=1:size(M,1)
    F(i) = sqrt((M(i,1)-x(1))^2+(M(i,2)-x(2))^2)-x(3);
end

options = optimset('LevenbergMarquardt','on');
x = lsqnonlin(@(x) MatlabNonLinearLSCircle(x,A),
    [0 0 0],[],[],options);
xc = x(1); yc = x(2); r = x(3);
[xc yc r]
```

17.2 Initial estimates for spheres:

```
function [xco,yco,zco,ro] = InitialEstimateLSSphere(M)
n = size(M,1);
for i=1:n
    A(i,1) = 2*M(i,1);
    A(i,2) = 2*M(i,2);
    A(i,3) = 2*M(i,3);
    A(i,4) = -1;
    B(i,1) = M(i,1)^2+M(i,2)^2+M(i,3)^2;
end
ATA = A'*A;
IATA = inv(ATA);
P = IATA*A'*B;
xco=P(1);
yco=P(2);
zco=P(3);
ro = sqrt(xco^2+yco^2+zco^2-P(4));
```

17.3 Gauss–Newton estimates for sphere:

```
function [xc,yc,zc,r] = LSSphereGN(M)
eps = 1e-8;
flag = 0;
n = size(M,1);
```

```
% Initial estimate
[xco,yco,zco,ro] = InitialEstimateLSSphere(M);
Eo = sum((sqrt( (M(:,1)-xco).^2+
(M(:,2)-yco).^2+ (M(:,3)-zco).^2) - ro).^2 );
t = 1;
while flag == 0
% Jacobian
    for i=1:n
        R = sqrt((M(i,1)-xco)^2+(M(i,2)-yco)^2+(M(i,3)-zco)^2);
        J(i,1) = -(M(i,1)-xco)/ro;J(i,2)=-(M(i,2)-yco)/ro;J(i,3)=
                -(M(i,3)-zco)/ro;J(i,4)=-1;
        D(i,1) = -(R-ro);
    end
    P = J\D;
    xco = xco + P(1); yco = yco + P(2); zco = zco + P(3) ;
    ro = ro + P(4);
    E1 = sum((sqrt((M(:,1)-xco).^2+(M(:,2)-yco).^2+(M(:,3)
         -zco).^2) -ro).^2);
    if abs(Eo-E1) < eps
        flag = 1;
    end
    Eo = E1;
    disp(['iteration # ' num2str([t xco yco zco ro Eo])]);
    t = t + 1;
end
xc = xco; yc = yco; zc = zco; r = ro;
```

17.4 The nominal center coordinates are $(1,1,-25)$ mm and the nominal radius is 25 mm. The actual values will depend on the random components in the data.

18.2 (a) The center coordinates are $(1.4529, 1.4529)$ and the radius is 18.6440. (b) $(1.4142, 1.4142)$ µm and 18.5858 µm for center and radius. In part (b), the data was simulated using a limacon approximation and also best-fit using a limacon approximation. Therefore the answers in part (b) appear to be correct, while those in part (a) appear to be incorrect.

18.3 Function to compute distances from cylinder:

```
% M is a nx3 matrix
% first column is the angle theta, second column is Z height
% and third column is radial deviation
function d = ComputeLSCylinderDistance(M,a,b,e,f,r)
n = size(M,2);
for i=1:n
    d(i,1) = M(i,3) - ((a+M(i,2)*e)*cos(M(i,1))
               + (b+M(i,2)*f)*sin(M(i,1)) + r);
end
```

19.1 One possible implementation for the brute force method is as follows. Given any three points whose indices are i, j, and k, three possible boundaries may be constructed, with i and j, or i and k or j and k on one boundary line and the remaining point on the other. The half width has to be determined and stored for each case. Also, a flag may be defined that contains information as to whether the current boundary offers a feasible solution or not (do all points lie inside the lines?). The indices i, j, and k are varied in a loop where i ranges from 1 to n, j from $i+1$ to n and k from $j+1$ to n. After the execution of the loop, only feasible solutions are searched to identify the set of three indices that correspond to smallest zone width.

19.2 The indices of the critical points are $(7,9,12)$. The MZ half-width h is 4.4627 μm. It takes several hundred exchanges for the brute force method and about five to seven exchanges depending on the starting solution (7 exchanges if point indices 1, 2, 3 are taken as the starting solution).

19.3 MZ half-width h is 0.1 mm. It takes about several thousand exchanges using the brute force as opposed to just nine with the alternating rule (with 1, 20, and 40 as starting indices).

19.4 The zone half-width is 1 μm. The MZ solution is not unique; different sets of four points can be determined that have the same MZ width.

20.1 Solving the dual LP problem in MATLAB.

```
theta = (0:7)'/8*2*pi;
Aeq = [cos(theta) sin(theta)]';
Beq = [0;0];
A = ones(1,8)
B = 1;
F = [1.1;1.4;1;1.6;1.3;1.2;1.5;1.4]'*10;
T=linprog(-F,A,B,Aeq,Beq,[0,0,0,0,0,0,0,0],
  [Inf Inf Inf Inf Inf Inf Inf Inf]);
F*T
```

20.2 The 180 Rule program. The indices of the current set of three defining points are p_1, p_2, and p_3. The outlier point has an index i.

```
function [p1,p2,p3] = oneeightyrule(p1,p2,p3,i,theta);
if (i < p1)
    if (theta(p2) - theta(i) > pi)
        p3 = p2;
        p2 = p1;
        p1 = i;
    else
        p1 = i;
    end
elseif (i < p2)
    if (theta(p3) - theta(i) > pi)
        p1 = i;
    else
        p2 = i;
    end
elseif (i < p3)
    if (theta(i) - theta(p1) > pi)
        p3 = i;
    else
        p2 = i;
    end
else
```

```
    if (theta(i) - theta(p2) > pi)
        p1 = p2;
        p2 = p3;
        p3 = i;
    else
        p3 = i;
    end
end
```

20.3 The limacon fit program is shown below. R is the array of radial deviations. The objective is to fit a limacon through three points whose indices are p_1, p_2, and p_3.

```
function [R,a,b] = limaconfit(R,p1,p2,p3,theta)
A=[1 cos(theta(p1)) sin(theta(p1));
    1 cos(theta(p2)) sin(theta(p2));
    1 cos(theta(p3)) sin(theta(p3))];
B=[R(p1); R(p2); R(p3)];
P = inv(A)*B;
R = P(1);
a = P(2);
b = P(3);
```

20.5 The primal can be written in the following standard form:

$$\begin{aligned}
&\text{Maximize } Z = -h\\
&\text{such that: } -(a_1-a_2)\cos\theta(i)-(b_1-b_2)\sin\theta(i)-r-h\leqslant -R(i)\,, \quad \text{for } i=1:n\\
&\qquad (a_1-a_2)\cos\theta(i)+(b_1-b_2)\sin\theta(i)+r-h\leqslant R(i)\,, \quad \text{for } i=1:n\\
&\qquad a_1,a_2,b_1,b_2,r,h\geqslant 0
\end{aligned}$$

Clearly, the origin is not a feasible solution. It does not satisfy the first set of n constraints. The corresponding dual is given by

$$\begin{aligned}
\text{Minimize } U = &-\sum_{i=1}^{n} u(i)R(i)+\sum_{i=1}^{n} v(i)R(i)\\
\text{such that } &-\sum_{i=1}^{n} u(i)\cos\theta(i)+\sum_{i=1}^{n} v(i)\cos\theta(i)=0\\
&-\sum_{i=1}^{n} u(i)\sin\theta(i)+\sum_{i=1}^{n} v(i)\sin\theta(i)=0\\
&-\sum_{i=1}^{n} u(i)+\sum_{i=1}^{n} v(i)\geqslant 0\\
&-\sum_{i=1}^{n} u(i)-\sum_{i=1}^{n} v(i)\geqslant -1\\
&u(i),v(i)\geqslant 0
\end{aligned}$$

In standard LP form, this translates to:

$$\begin{aligned}
\text{Maximize } U = & \sum_{i=1}^{n} u(i)R(i) - \sum_{i=1}^{n} v(i)R(i) \\
\text{such that } & \sum_{i=1}^{n} u(i)\cos\theta(i) - \sum_{i=1}^{n} v(i)\cos\theta(i) = 0 \\
& \sum_{i=1}^{n} u(i)\sin\theta(i) - \sum_{i=1}^{n} v(i)\sin\theta(i) = 0 \\
& \sum_{i=1}^{n} u(i) - \sum_{i=1}^{n} v(i) = 0 \\
& \sum_{i=1}^{n} u(i) + \sum_{i=1}^{n} v(i) \leqslant 1 \\
& u(i), v(i) \geqslant 0
\end{aligned}$$

Note that there are $2n$ dual variables. We have used the notation u for the set of n variables that correspond to the first set of n primal constraints, and v for the second set of n primal constraints. Clearly, each $u(i) = v(i) = 0$ satisfies the constraints, and therefore the origin is a feasible solution.

20.6 The zone width $2h$ is 4.8284 μm.

20.8 There are 16 elements in T (the output of `linprog`) and these represent the values of the 16 dual variables. The first eight dual variables correspond to the first eight primal constraints, which in turn correspond to the outer circle. Of these, only dual variables 2 and 4 are non-zero. From the complementary slackness theorem, slack variables corresponding to constraints 2 and 4 are zero, which implies that points 2 and 4 contact the outer circle. The second eight dual variables correspond to the inner circle. $T(11)$ and $T(14)$ are non-zero and these correspond to constraints 3 and 6 for the inner circle whose slack variables are zero. Therefore, points 2 and 4 lie on the outer circle and 3 and 6 lie on the inner circle. The `mzcircle` program verifies this: indices of points are $p_1 = 2$, $p_2 = 3$, $p_3 = 4$, and $p_4 = 6$, with p_1 and p_3 on the outer circle and p_2 and p_4 on the inner circle. Note that positive sign for h implies this; negative h would imply p_1 and p_3 on the inner circle.

20.9
(a) The solution is $a = 0.9999962$ μm, $e = -0.999989$ μm/mm, $b = -1.0860835$ μm, $f = 0.543038$ μm/mm, and $r = 10.000004$ μm.
(b) The radius of the MC cylinder is 15.8153 μm.
(c) The zone width $2h = 6.5858$ μm.
(d) The solution obtained in (c) only provides the values for the dual parameters. However, as discussed in the solution for Exercise 20.8, we can identify the set of points that lie on the MZ cylinder boundaries (using complementary slackness theorem). In the primal form, we can solve for the set of unknowns using just those points that lie on the cylinder. This would be a simultaneous solution for six unknowns using six equations.

21.2 *Ra*, *Rq*, *Wa*, and *Wq* as function of cutoff:

λ (mm)	*Ra* (μm)	*Rq* (μm)	*Wa* (μm)	*Wq* (μm)
0.25	0.0070	0.0078	0.6472	0.7142
0.8	0.0646	0.0724	0.5512	0.6177
2.5	0.4211	0.4677	0.2156	0.2394

21.3 Yes, $RSm = 2$ mm for the profile in this exercise. It will always be 2 mm regardless of phase offsets. The phase offset only serves to reduce the number of acceptable intervals that can be found. The spacing between acceptable zero crossings is always 2 mm.

21.4 The slope values are 87.4216° and 47.9921°; z_1 has a larger amplitude at smaller wavelengths compared to z_2, and is therefore expected to have larger slope.

21.5 Kurtosis of the random profile is close to 3. Kurtosis of the other two profiles are 1.8 and 5.8. The profile in part (c) is spikiest of all and has the largest kurtosis.

22.2 A function to compute unbiased right and left-side *ACV*:

```
function [ACV, ACF, shift] = computeUnbiasedACF(z,dx)
n = size(z,1);                             % length of profile z
Rq = sqrt(sum(z.*z)/n);                    % determine Rq
for k = 0:n-1
    temp = 0;
    for j=1:n-k
        temp = temp + z(j,1)*z(j+k,1);     % compute the sum
                                           % of products
    end
    ACV_right(k+1,1) = temp/(n-k);         % divide by number
                                           % of points
end
for k = -n+1:1:-1
    temp = 0;
    for j=1-k:n
        temp = temp + z(j,1)*z(j+k,1);     % compute the sum
                                           % of products
    end
    ACV_left(k+n,1) = temp/(n+k);          % divide by number
                                           % of points
end
ACV = [ACV_left;ACV_right];
shift = (-n+1:n-1)'*dx;                    % generate the x axis
                                           % array of shift
                                           % distances

subplot(2,1,1),plot(shift,ACV);            % plot ACV
xlabel('Shift (mm)');
ylabel('ACV');
ACF = ACV/Rq^2;
subplot(2,1,2),plot(shift,ACF);            % plot ACF
xlabel('Shift (mm)');
ylabel('ACF');
```

22.3 The weighted sum can be performed using MATLAB's `conv` function. Consider one copy of z_1 as the profile and another copy as the weighting function. Recall that `conv` first flips the weighting function prior to performing convolution. We therefore flip the weighting function prior to sending it through `conv`.

```
C = conv(z1,flipud(z1));
plot(shift,C,'r-')
```

Notice that the weighted sum closely resembles the left- and right-side-biased *ACV* in shape but not magnitude. However, the definition of the *ACV* is itself a moving average, but scaled down by a $(1/n)$ term. Therefore, $C/8{,}000$ is identical to the biased left- and right-side *ACV*. In the function written in Exercise 22.2, replace lines

```
ACV_right(k+1,1) = temp/(n-k);
ACV_left(k+n,1) = temp/(n+k);
```

by

```
ACV_right(k+1,1) = temp/(n)
ACV_left(k+n,1) = temp/(n);
```

Superimpose the resulting *ACV* on $C/8{,}000$. They should be identical.

22.4 Evaluating the integral, we get

$$zf(f) = \int_{x=0}^{x=l} \sin(2\pi f_o x) e^{-i2\pi f x} dx =$$

$$\int_{x=0}^{x=l} \sin(2\pi f_o x)\cos(2\pi f x) dx - i \int_{x=0}^{x=l} \sin(2\pi f_o x)\sin(2\pi f x) dx \,.$$

Because f_o and f are spatial frequencies defined as (m_1/l) and (m_2/l), m_1 being a constant integer and m_2 any integer, the integrals to evaluate can be expressed, by replacing $\omega = 2\pi x/l$, as

$$zf(f) = \frac{l}{2\pi}\left[\int_{\omega=0}^{\omega=2\pi} \sin(\omega m_1)\cos(\omega m_2) d\omega - i \int_{\omega=0}^{\omega=2\pi} \sin(\omega m_1)\sin(\omega m_2) d\omega\right] .$$

The first integral is always zero. The second integral is non-zero only when $m_1 = m_2$. Therefore,

$$zf(f_o) = -\frac{l}{2\pi} i\pi = -il/2 \,.$$

Therefore,

$$|zf(f_o)|^2 = \frac{l^2}{4} \quad \text{and } PSD = l/4 \,.$$

22.5 For load-carrying purposes, it is desirable to have a higher concentration of points near the top of the surface so that the load can be distributed evenly. A plot of the histogram of each of the surfaces will indicate where the points are distributed.

```
[H1,binLocation1] = ADF(z2, 50); % determine ADF using 50 bins
plot(H1,binLocation1);
xlabel('Frequency (#)');
ylabel('Bin Location (\mum)');
[H1,binLocation1] = ADF(z3, 50); % determine ADF using 50 bins
plot(H1,binLocation1);
xlabel('Frequency (#)');
ylabel('Bin Location (\mum)');
```

The histogram reveals that profile z_2 has a higher concentration of points at the top, and is therefore more suitable.

23.2 *Sku* is approximately 3. *Ssk* is approximately 0. The surface does not have any skew (hence a value of zero) and is normally distributed (hence a value of 3).

23.3 The surface contains two sinusoids, one along x and another along y, each of wavelength 6.4 mm. The *APSD* matrix will therefore have a peak at $(1,3)$ and $(3,1)$. Note that $(1,1)$ is the DC term. $(1,2)$ represents a surface wavelength of $l/1 = 12.8$ mm along x, and $(1,3)$ represents a surface wavelength of $l/2 = 6.4$ mm along x. There will also be peaks at the mirror position $(1,127)$ and $(127,1)$, which represents -6.4 mm.

23.4 Because of the periodicity along x and y, the *AACV* rises and falls in both directions. Note the period along x and y is 6.4 mm.

23.5 Note that the histogram does indeed resemble a normal distribution, indicating the surface is Gaussian.

25.1 The mean value of *Ra* is 0.2598 μm (evaluated on the filtered profile between 0.8 and 7.2 mm) and the standard deviation is 0.00056 μm and the interval corresponding to 95% level of confidence is [0.2587–0.2609] μm.

25.2 The mean value of *Rt* is 1.6190 μm and the standard deviation is 0.0496 μm. The uncertainty is larger because *Rt* depends only on two points, while *Ra* is averaged over many more points. The 95% level of confidence is [1.5351–1.7306] μm.

25.3 According to the law of propagation of uncertainty, the uncertainty in the LS center and radius can be expressed as

$$
\begin{aligned}
u(r) &= \sqrt{\frac{u^2(R(1))}{n^2} + \ldots \frac{u^2(R(n))}{n^2}} = \frac{0.1}{\sqrt{n}} = 0.005\,\mu\text{m}\,. \\
u(a) &= \sqrt{u^2(R(1))\cos^2\theta(1)\frac{4}{n^2} + \ldots + u^2(R(n))\cos^2\theta(n)\frac{4}{n^2}} = \\
&\quad 0.1\sqrt{1/180} = 0.0075\,\mu\text{m}\,. \\
u(b) &= u(a) = 0.0075\,\mu\text{m}\,.
\end{aligned}
$$

Note that $\sum_{i=1}^{n} \cos^2\theta(i) = \sum_{i=1}^{n} \sin^2\theta(i) = n/2$.

25.4 The functional relationship between the residual profile points and the LS circle can be written as $d(\theta) = R(\theta) - (r + a\cos\theta + b\sin\theta)$. Because r, a, and b are obtained from the original profile R, there is expected to be some correlation between these input parameters. Ignoring the correlation terms, the uncertainty in the output can be expressed as

$$u(d(\theta(i))) = \sqrt{u^2(R(\theta(i))) + u^2(r) + u^2(a)\cos^2\theta(i) + u^2(b)\sin^2\theta(i)} = 0.1004\,\mu\text{m}.$$

From MCS, the uncertainty of any point (the average of the uncertainties at each of the 360 points as the representative value) is equal to 0.0996 μm. This value is somewhat smaller than what we computed using GUM, indicating that correlation terms, although small, have to be accounted for. The uncertainty in out-of-roundness is 0.065 μm.

26.2 The three-probe method is sensitive to the 3rd harmonic (can reconstruct) but not sensitive to the 35th (cannot reconstruct) for a probe angle of 10°.

26.3 Yes, the reconstruction succeeds because the transfer function does not drop to zero.

26.4 There is never a harmonic suppression. The limitation of this method is that the probe spacing has to be equal to the sampling interval, and because finite probe size limits how close they can be placed to each other, the sampling interval is not as small as desirable.

26.5 Notice that the first three zeros occur at the 36th, 72nd, and 108th harmonics as expected. Because the magnitude of Hf drops to zero for those harmonics which satisfy $\omega = 360n_1/\phi_1 = 360n_2/\phi_2$, where n_1 and n_2 are any integers, if we choose $\phi_1 = 10°$ and $\phi_2 = 15°$, the first common zero occurs only at 72° ($n_1 = 2, n_2 = 3$).

26.6 A function to generate three-position data:

```
function [theta,R,m1,m2,m3,n,phi,s] =
genDataThreePosition(n,phi, partUPR)
% n is the number of points measured by 1 probe in 1 revolution
% phi is the angle between probes in degrees
% part error
theta = (0:1:n-1)'*2*pi/n;
R = sin(theta*partUPR);
R = R - mean(R);
% spindle error
k = (1:n)';
s = 0.02*sin(2*pi.*k/n*12);
s = s - mean(s);
% angle between probes
t = 360/n; % t is the sampling interval in degrees
q = phi/t; % the probe is located at every qth point
phi = phi*2*pi/360;
% the data that the probes see
m1 = R + s;
R1 = [R(q+1:n); R(1:q)];
m2 = R1 + s;
```

```
R2 = [R(2*q+1:n); R(1:2*q)];
m3 = R2 + s;
```

26.7 (a) The three position FFT problem can be solved as shown:

```
n = 36;
[theta,R,m1,m2,m3,n,phi] = genDataThreePosition(n,10,3);
phi = 10 *2*pi/360;
M = 2*m1 - m2 -m3;
Mf = fft(M);
w = 0:1:n/2+1;
H = 2 - exp(j*w*phi) - exp(j*2*w*phi);
for i=1:n/2+1
    if abs(H(i)) > 1e-8
        R1(i) = Mf(i)/H(i);
    else
        R1(i) = 0;
    end
end
R1 = [R1 fliplr(conj(R1(2:n/2)))];
R1 = real(ifft(R1));
polar(theta,R+5,'k*'); hold on; polar(theta,R1'+5,'r-')
```

(b) The problem may also be formulated as a set of linear equations as shown below:

```
ProbeAngle = 10;
[theta,R,ma,mb,mc,n,phi] = genDataThreePosition(36,ProbeAngle,3);
Y = [ma;mb;mc];
Rc = diag(ones(1,36)); % Rc represents the coefficients for the
                       % reconstructed part form matrix
S = diag(ones(1,36));  % S represents coefficients for the
                       % spindle error matrix
% Rc1 is the matrix of coefficients for all three positions. The
% first 36 rows correspond to the first position, the next 36
% correspond to the next position and the final 36 rows to the
% last position.
Rc1 = [Rc;Rc(2:36,:);Rc(1,:); Rc(3:36,:);Rc(1:2,:)];
S1 = [S;S;S];
U = [Rc1 S1];
rank(U)
```

Notice that the rank of U is 71. We therefore need one additional constraint. Arbitrarily setting the average spindle error to be zero, we have

```
U(109,1:36) = 0; U(109,37:72) = 1;
Y(109,1) = 0;
```

Verify now that rank(U) is 72. Solving for V,

```
V = U\Y;
polar(theta,R+5,'g*'); hold on; polar(theta,V(1:36)+5,'r-')
```

26.8 From GUM, the standard uncertainty at any point in the reconstructed profile is $\frac{u(m(\theta))}{\sqrt{2}}$, where the uncertainty in each measured point of the profile is given by

$u(m(\theta))$. Thus, the uncertainty is $0.1/\sqrt{2} = 0.07\,\mu\text{m}$. This can be verified using MCS as shown below. In the code segment below, *R* is the part form, *S* is the systematic component of the spindle error, m_1 and m_2 are the measured profiles, u_1 and u_2 are the uncertainty terms for each measured point drawn from the corresponding normal distributions and *Rc* is the reconstructed part form error. There are 360 sampling points along the circumference and each point is inspected 5,000 times for MCS routine. The standard deviation at each of the sampling points may not be identical; the average standard deviation (or pooled standard deviation) will be a reasonable estimate for the uncertainty of any point in *Rc*.

```
n = 360;
theta = (0:n-1)'/n*2*pi;
R = sin(3*theta);
S = rand(360,1)*0.3;
m1 = R + S;
m2 = R - S;
for i=1:5000
    u1 = normrnd(0,0.1*ones(360,1));
    u2 = normrnd(0,0.1*ones(360,1));
    Rc(i,:) = ((m1+u1) + (m2+u2))/2;
end
mean(std(Rc,0,1))
```

Index

MIX
Papier aus verantwortungsvollen Quellen
Paper from responsible sources
FSC® C105338

If you have any concerns about our products,
you can contact us on
ProductSafety@springernature.com

In case Publisher is established outside the EU,
the EU authorized representative is:
Springer Nature Customer Service Center GmbH
Europaplatz 3, 69115 Heidelberg, Germany

Printed by Libri Plureos GmbH
in Hamburg, Germany